BRITAINS CIVILIAN TOY FIGURES

NORMAN JOPLIN

ARNOLD ROLAK

PHILIP DEAN

JOE KUNZELMANN

Schiffer Publishing Ltd

4880 Lower Valley Road, Atglen, PA 19310 USA

DEDICATION

For Sheila – always.

To the memory of Anne Benge, a dear and loyal friend, and David Clarke, a collector and toy soldier dealer whose friendship will always be missed. Both, sadly, are no longer with us.

Library of Congress Cataloging-in-Publication Data

Britains civilian toy figures / by Norman Joplin...[et al.].
p. cm.
ISBN 0-7643-1520-X
1. Britains Ltd.--Catalogs. 2. Miniature objects--Great Britain--Collectors and collecting--Catalogs. I. Joplin, Norman.
NK8473.5.B75 A4 2002
688.7'2--dc21
2001006597

Designed by Bonnie M. Hensley
Cover design by Bruce M. Waters
Type set in BankGothic Md BT/Korinna BT

ISBN: 0-7643-1520-X
Printed in China
1 2 3 4

Published by Schiffer Publishing Ltd.
4880 Lower Valley Road
Atglen, PA 19310
Phone: (610) 593-1777; Fax: (610) 593-2002
E-mail: Schifferbk@aol.com
Please visit our web site catalog at **www.schifferbooks.com**
We are always looking for people to write books on new and related subjects. If you have an idea for a book please contact us at the above address.

This book may be purchased from the publisher.
Include $3.95 for shipping.
Please try your bookstore first.
You may write for a free catalog.

In Europe, Schiffer books are distributed by
Bushwood Books
6 Marksbury Ave.
Kew Gardens
Surrey TW9 4JF England
Phone: 44 (0) 20 8392-8585; Fax: 44 (0) 20 8392-9876
E-mail: Bushwd@aol.com
Free postage in the U.K., Europe; air mail at cost.

CONTENTS

ACKNOWLEDGMENTS

Daniel Agnew	Christie's Auction House, South Kensington, London
Craig Anderson	Choice Photo, Green Valley, Arizona
Bob Bailey	Access to Collection for Photography
Debbie Bednarek	Britains Garden
Britains/Racing Champions	For permission to use the Britains name, logo and trade mark
Dudley Chignall	The entire Cococub chapter research
Richard Charlesworth	Britains Footballers
Doug Congdon-Martin	Senior Editor, Schiffer Publishing
Philip Dean	Photography in the United Kingdom and Co-Authorship
Rex George	Britains Farm
Ray Haradin	Rare and unusual items in most chapters
Anne and Brian Holmes	Britains horse drawn vehicles, Farm and Hunt
Peter Kirk	The entire Racing Colors chapter research
Joe Kunzelmann	Britains Farm, Zoo and Circus
Hugo Marsh	Christie's Auction House, South Kensington, London
Scott Morlan	Civilian items in general
Alf Mussett	Britains Footballers research
David Pressland, New Cavendish Books	For permission to use photographs from several of their published books
Colonel Donald Pudney	Britains Factory records research
Arnold Rolak	Co-Authorship and Britains Factory records research
Carol Runkle	Typing the manuscript and, above all, the friendship
Stewart Saxe	Rare Cowboys and Indians
Hank and Doris Sigourney	Salvation Army
Steve and Josie Sommers	For permission to use material from *Old Toy Soldier Magazine*
Deborah Strutt	Freelance photography
Peter Schiffer and the Staff at Schiffer Publishing	
Vectis Auctions, Ltd., England	
Joe Wallis	For sharing his vast knowledge of Britains products
Betty Wheeler	For the Brown World Cow Hunt
Thomas and Catherine Wiedersiem	
Paul Williams	Specialist photographic processing
Ray Williams	R.A.C. figure
Dr. Robert Wilson	For his expert advice in the preparation and structuring of the book
Kent Wood	Photography in the U.S.A. and Canada
Ray Wooton	Britains, Ltd./Racing Champions
Robin Yates	Australian research into Britains and Department Store information

In addition, the preceding also contributed much enthusiasm and support through their loyalty and friendship. They were joined by many good friends who, although not contributing directly to the book, provided vital support during some of the difficult times. To all, I extend my sincere thanks: Dawn Allen, Ron and Debbie Eccles, Lawrence Fabacher, Dr. William Gray, Sylvia Hart, Verne Johnson, Don Pielin, Lou Sandbote, Steve White, Jacqueline Wilson, and Dana Wood.

1. Introduction

Our purpose is to issue a comprehensive reference book on the non-military products of Britains Limited between 1880 and 1966. While the *Great Book of Hollow Cast Figures* has previously been considered the definitive reference work on the subject, that book covered only a small proportion of the items available from Britains, as it concentrated on the other companies that were Britains competitors. In depth is probably the phrase that should be used to describe the content of this book, with one of the unique features being the previously unpublished records from the Britains archive built up jointly by Arnold Rolak and Colonel Donald Pudney. These records give intricate details of Master Model and Master Mold numbers. Much of this information has never before been published and is enhanced by artist impression drawings of items not produced. This book should serve as the most comprehensive ever written, as color variants for as many items as possible have been included in the illustrations.

Another unique facet of the information in the book is that for the first time an attempt has been made to give every single non-military Britains' item its own unique identity number, an example being the Boy Scout range. Issued in boxed sets, up to now the individual items that made up these sets have not had their own identity and have, therefore, only been referred to by the number of the set that they appeared in. The Scout Master was allocated Master Model #15B (see explanation of numbering system) and, while some of these numbers have emerged over the years in various publications and in the Britains Archive auction held at Christie's in London, 1994, no concise record has been available. In the absence of a catalog number, set number, Master Model number, it has been the task of the authors to come up with an identity for such figures and with the availability of the Master Mold list, this, in most cases, has been achieved. So, for the first time, each individual figure can at last be referred to with its own unique identity number.

Minor changes to some products would often result in a change of number, i.e., the color of the Eastern shepherd man's beard resulted in two Master Model #3s to distinguish between the two items. Items included in a standard cataloged set, but not part of the main title of the set, needed their own identity. For instance, there is a set number for the Royal Mail Van, but it is now possible to allocate a number to the driver with that number being different from the driver of the Volunteer Ambulance.

Criteria

The following criteria was used when deciding in which order the book should be organized.

Starting in 1880 and working through the production history of the Britains company, each chapter has been presented in the order that the first example from that series or range was first introduced into the Catalog, thus, enabling the collector to track the progress as each range was developed. The use of the phrases pre-War or post-War refer to World War Two.

Chapter names include the starting dates in each range. Some of the dates may be approximate, as some of the companies' earlier records have not yet been found. Some ranges may have been introduced only briefly and then reintroduced in larger quantities at a later date.

Explanation of Numbering Systems

Unless otherwise stated, all numbers (#) are catalog numbers. Some catalog numbers have "B" suffix indicating that instead of allocating a catalog number, the Master Model number was used. In some, but not all of these instances, the "B" suffix indicated "Bulk Lines." These refer to quantities of unboxed items issued to the retail trade. The Picture Pack Series also utilized this "B" suffix number as a catalog identity.

Up until 1994, when the Britain's Archive was auctioned in London by Christie's, it was not generally known outside of the Britains Company why the "B" suffix appeared in the catalog against certain lines instead of the conventional catalog number.

The Archive contained many sample cards used for identifying individual figures within the factory. Each figure was tied to a card in a "rough" numerical order system. These numbers all ended with the letter "B." Not all the figures produced, nor all the sample cards, remained in the Archive, thus leaving many items still without an individual identity.

Arnold Rolak and Donald Pudney obtained a copy of the Britains Factory records and it was at this point that it was discovered that the "B" suffix numbers were in fact the Master Model numbers used by Britains. Many figures did not have an individual identity if contained and then removed from a boxed set. Example: Scout Master issued in various boxed sets is identified as a single figure by the Master Model #15B.

Master Mold Numbers

The factory records also contained lists of the molds held within the Britains Factory. Each mold was numbered for identification purposes. In instances where no catalog or Master Model number is available, the Master Mold number has been indicated. Master Mold numbers do not have a "B" suffix or are related in any way to catalog numbers.

If a full set is illustrated, the set number from the catalog numbering system is used.

Price Guide

The Price Guide is based on author Norman Joplin's 30 year experience in the toy figure hobby and is intended only to be a guide for collectors.

The condition of figures is a factor that is very important and the guide has been compiled on the assumption that each item is in good to very good condition in relation to paint. A plus or minus must be applied to each item depending on better or worse condition. Structural damage to figures and figures that have been repaired or repainted lose much of their value if they are being acquired for investment. There are no definitive solutions to the price of any items, as it is a matter between vendor and purchaser to determine the value.

Items considered and indicated as "unique" have not been given a value.

Terminology

English	American
Automobile Association (AA)	AAA
Dungarees	Bib Overalls
Football	Soccer
Kennel	Dog House
Hunt	Fox Hunting
Lorry	Truck
Petrol Pumps	Gas Pumps
Range	Line
Roundabout	Circle
Royal Automobile Club (RAC)	AAA
Waistcoat	Vest

Other Terms

Air Raid Warden: Civilian employed during World War II to assist during bombing attacks in the Blitz
Coster: A street vendor usually with a hand cart
Flint Wall: Flint is a kind of quarry stone used for building.
Navvy: A manual worker in construction
Royal Mail: The British mail service
Royal National Lifeboat Institute (RNLI): Volunteer civilian rescue crews.
St. John's Ambulance Brigade: Civil volunteer medical association.
Stoker: Railway worker who shovels coal in steam railway engines.
Strip: the overall garb, i.e. shirt, shorts, socks, shoes, etc., of a soccer player

Animal Breeds and the areas in the United Kingdom where they are bred.

Ayrshire (cattle): Scotland
Berkshire (pigs): England
Clydesdale (horses): England
Exmoor (sheep): England
Hampshire (sheep): England
Jersey (cattle): Channel Islands
Suffolk (horses): England

Spelling

English	American
Colour	Color
Catalogue	Catalog

Opposite page: The Edwardian Family. This set was reputedly based on the family of Frederick Britain and is thought to date from around 1907. Although a number of individual items from this set have been discovered by collectors from time to time, this is the only complete group known to exist. *Courtesy Christie's Auction House, London/The Arnold Rolak Reference Collection.* $10,000-12,000.

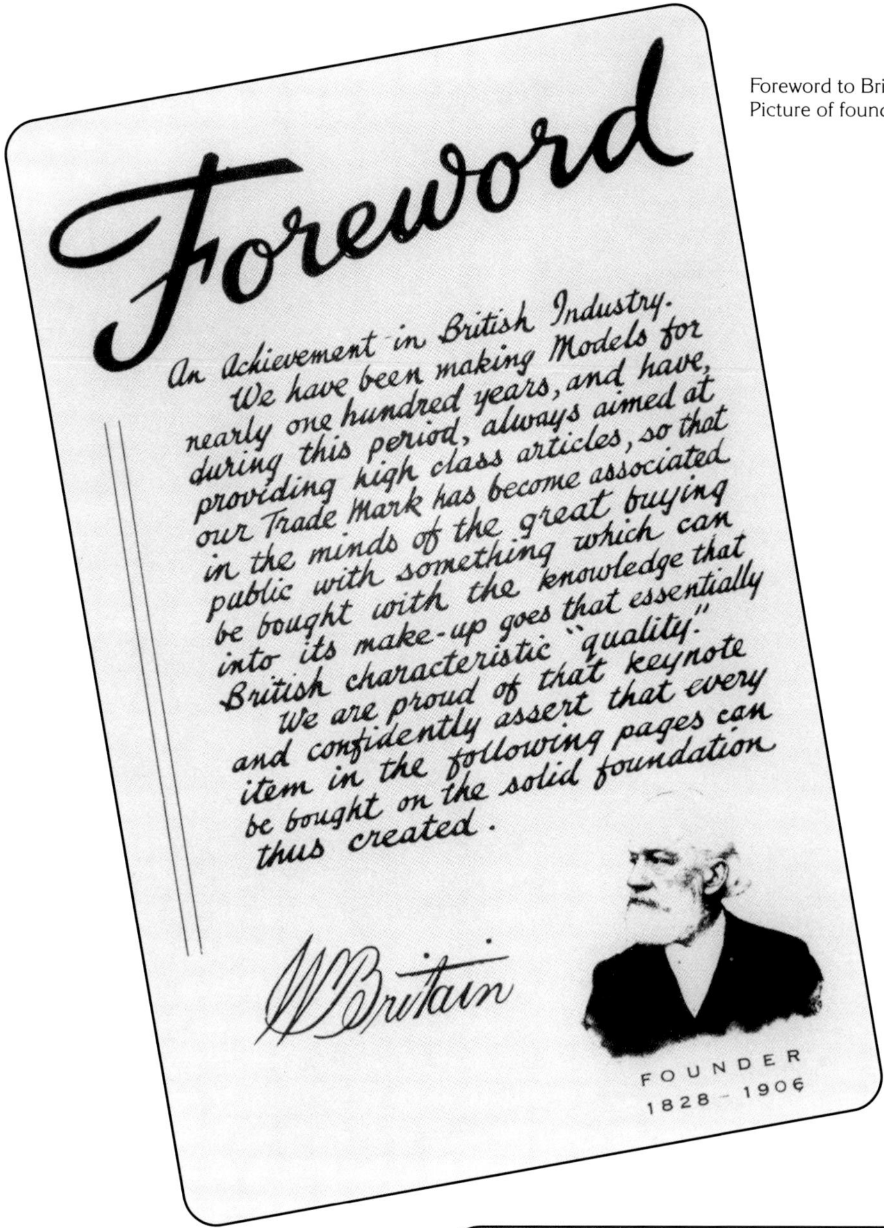
Foreword

An Achievement in British Industry.

We have been making Models for nearly one hundred years, and have, during this period, always aimed at providing high class articles, so that our Trade Mark has become associated in the minds of the great buying public with something which can be bought with the knowledge that into its make-up goes that essentially British characteristic "quality."

We are proud of that keynote and confidently assert that every item in the following pages can be bought on the solid foundation thus created.

W Britain

FOUNDER
1828 – 1906

Foreword to Britains Catalog.
Picture of founder.

Britains Ltd

January, 1953
108th Edition

This Catalogue contains a restricted range of lines, which will be added to as circumstances permit.

All Lines in this Catalogue are Hand-produced by highly trained British Craftsmen

Britains Logo.

MANUFACTURED BY W Britain IN LONDON, ENGLAND

TRADE MARK.

REGD. No. 459993.

Britains Trademark.

2. The 1880 Catalog and Beyond

Automatic, Scientific and Clockwork (Mechanical) Toys

The discovery of a catalog entitled *W. Britain and Sons* celebrated automatic and scientific clockwork models, toys, etc., believed issued in 1880 is the inspiration for this chapter. Arnold Rolak, who accumulated a unique Britains reference collection, has specialized in acquiring the contents of this catalog. Although not succeeding in tracking down every example, he came close to his goal before the auction of his collection at Christie's London in 2000. Marcus Grien, well-known expert on toy figure production in Germany, and Dr. Rob Wilson, who having carried out extensive research into the early products of Britains, now believe that the common assumption that Britains invented the hollow cast method of toy soldier production may be incorrect and that it was, in fact, a German invention. There is no doubt that many of the items issued by Britains and illustrated in the 1880 catalog were manufactured in Germany. Some examples have been found in boxes with printed labels in German and a German toy manufacturer's catalog in the possession of Marcus Grien shows examples of items not dissimilar to those in the Britains catalog. Dr. Rob Wilson is currently researching for a future reference book covering the Britains Paris office issues and it is hoped that as well as covering special paint and special orders of the nineteen thirties, information will be found verifying the German/Britains connection. The sequence in which illustrations are presented follow the same order as the 1880 catalog.

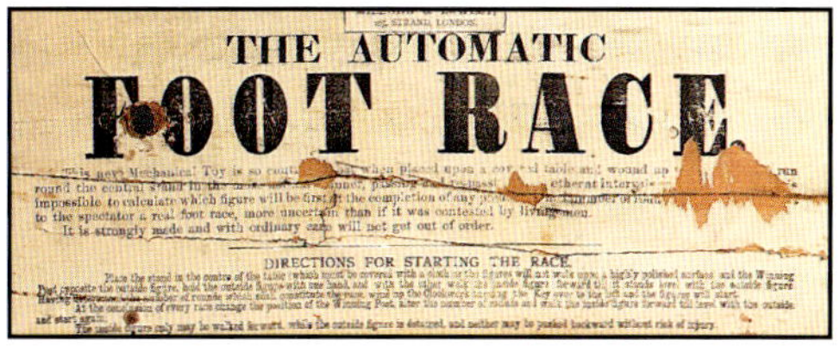

Top center: The Mechanical Foot Race. Close up of the walking men. *Courtesy Christie's Auction House, London/The Arnold Rolak Reference Collection. Specialty photo courtesy Philip Dean.*

Top right: Wooden box for the Mechanical Foot Race. *Courtesy Christie's Auction House, London/The Arnold Rolak Reference Collection.*

Bottom right: The Mechanical Foot Race. Two nine-inch high men dressed in silk clothing described as "walking match" in the catalog of 1880. *Courtesy Christie's Auction House, London/The Arnold Rolak Reference Collection.* $10,000-$12,000, including box.

The Mechanical Chinaman. A Chinese Mandarin in satin, twelve inches high. Nicknamed "The Drinking Chinaman" he raises his cup to drink his tea at the same time throwing his head back with satisfaction. *Courtesy Ray Haradin.* $8,000-$9,000.

The Walking Elephant. When the rider's parasol is spun, this operates the integral flywheel mechanism and the elephant walks. *Courtesy Christie's Auction House, London/ The Arnold Rolak Reference Collection.* $2,000-$2,500.

The Model Crane. *Courtesy Christie's Auction House, London/ The Arnold Rolak Reference Collection.* $150-$200.

The Miniature Road Roller with original box. Both this and the previous item are made to propel by spinning the flywheel. *Courtesy Christie's Auction House, London/The Arnold Rolak Reference Collection.* $1,000-$1,200.

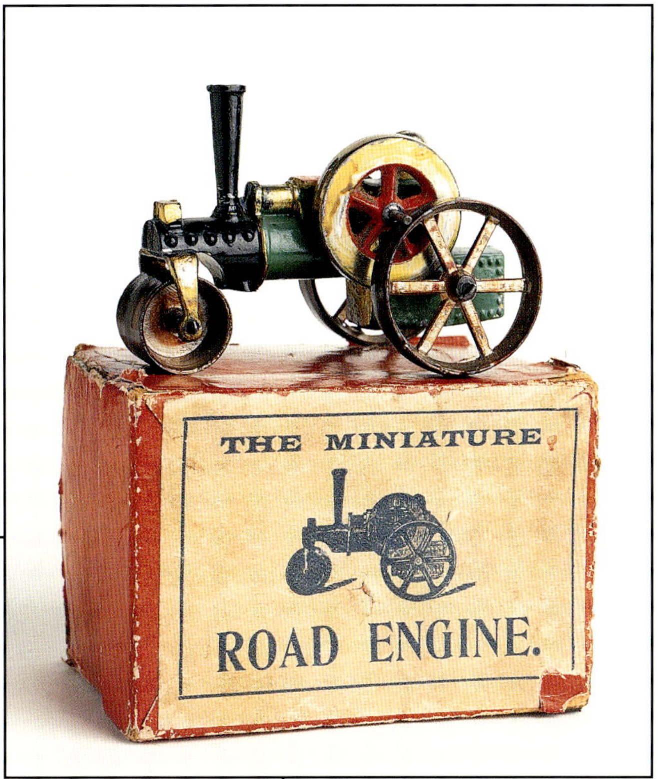

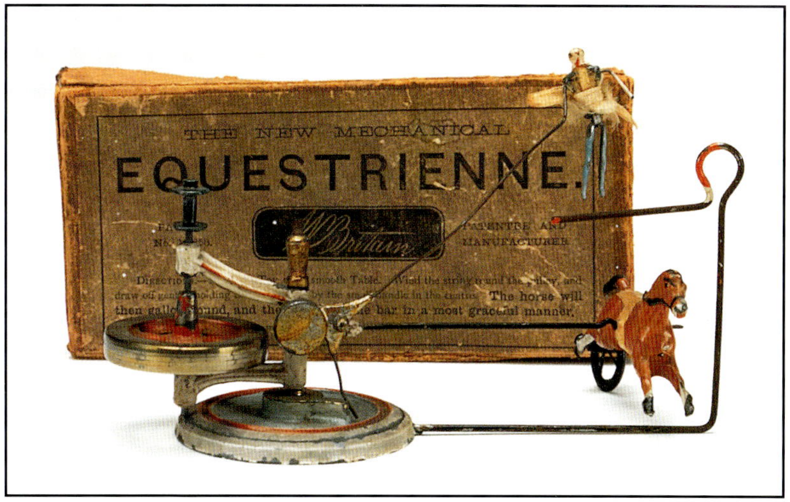

The Equestrienne with original box. As the horse rotates, the rider "jumps" the post a dual flywheel is employed. *Courtesy Christie's Auction House, London/The Arnold Rolak Reference Collection.* $400-$500.

The London Road Roller with original box. *Courtesy Christie's Auction House, London/The Arnold Rolak Reference Collection.* $2,200-$2,500.

Don Quixote and the Windmill with original box. The flywheel mechanism rotates Don Quixote around the windmill. Each of the windmill's fans is a different color, thus presenting the opportunity to participate in the game by guessing the color that the rider's lance will touch. *Courtesy Christie's Auction House, London/The Arnold Rolak Reference Collection.* $6,000-$7,000.

The Fountain Top with original printing plate for box artwork. The top is spun in the normal way; placed in a saucer of water it throws up a jet of water. *Courtesy Christie's Auction House, London/The Arnold Rolak Reference Collection.* $800-$900.

The Boomerang Top with original box. By spinning the top and placing "flyers" on top, the flyers are propelled into the air in random fashion. *Courtesy Christie's Auction House, London/The Arnold Rolak Reference Collection.* $1,200-$1,400.

The Mikado with original box. By spinning the parasol, the internal flywheel mechanism moves the hand containing a paper fan. *Courtesy Christie's Auction House, London/The Arnold Rolak Reference Collection.* $1,500-$1,600.

The Mikado with original box printed in German. See introduction to this chapter regarding the German/Britains Connection. *Courtesy Christie's Auction House, London/ The Arnold Rolak Reference Collection*. $700-$800.

Two Horse Race with the printing block for the original box art. See previous illustration. *Courtesy Christie's Auction House, London/The Arnold Rolak Reference Collection*. $1,300-$1,500.

The Clown. A plate balanced on the Clown's head, when spun, operates the flywheel mechanism to move the hand which holds a stick that spins the plate (stick missing). *Courtesy Christie's Auction House, London/The Arnold Rolak Reference Collection*. $1,000-$1,200.

The Four Horse Race. Flywheel driven horse race. *Courtesy Christie's Auction House, London/The Arnold Rolak Reference Collection*. $2,000-$3,000.

The Waltzing Couple with original box. A soldier and his lady waltzing. When spun, the couple revolve as if dancing. *Courtesy Christie's Auction House, London/The Arnold Rolak Reference Collection*. $2,800-$3,200.

The General with original box. Soldier mounted on a hobby horse and carrying an umbrella which, when spun, propels the horse. At the same time the soldier's sword arm moves to and fro. A blue-jacketed version also exists. The sample illustrated differs from the 1880 catalog illustration/line drawing in that he wears a tricorn hat rather than a German helmet. *Courtesy Christie's Auction House, London/The Arnold Rolak Reference Collection*. $6,000-$6,500.

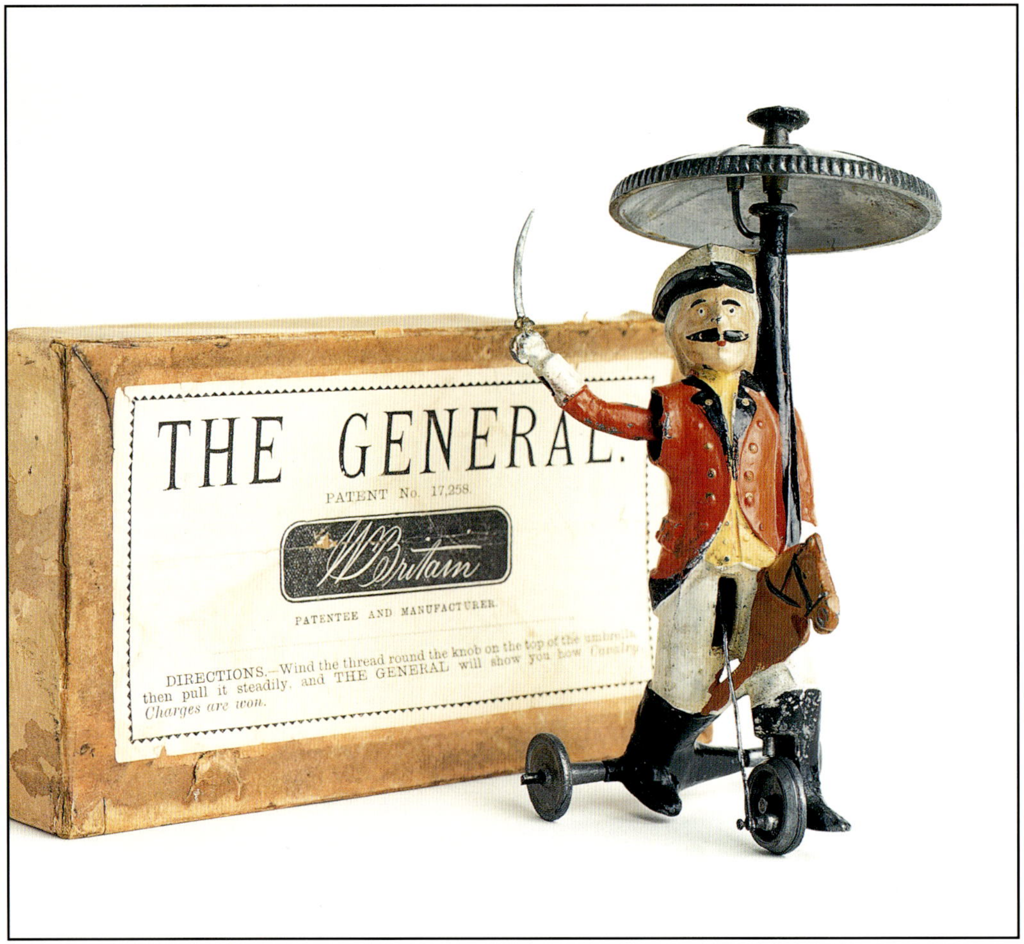

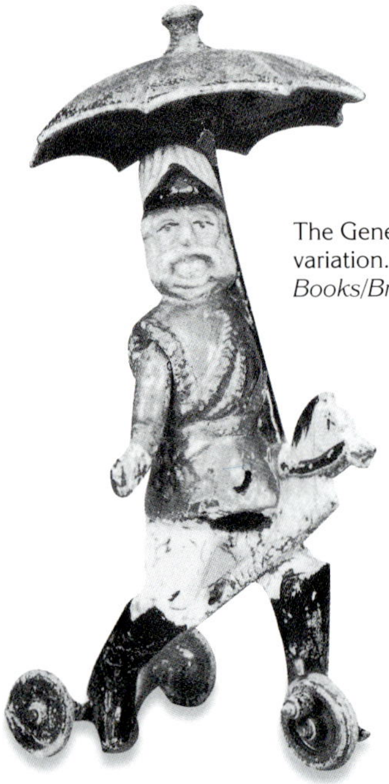

The General, showing a different variation. *Courtesy New Cavendish Books/British Tin Toys*. Unique.

Some of the following do not appear in the 1880 Catalog, but were probably produced shortly after.

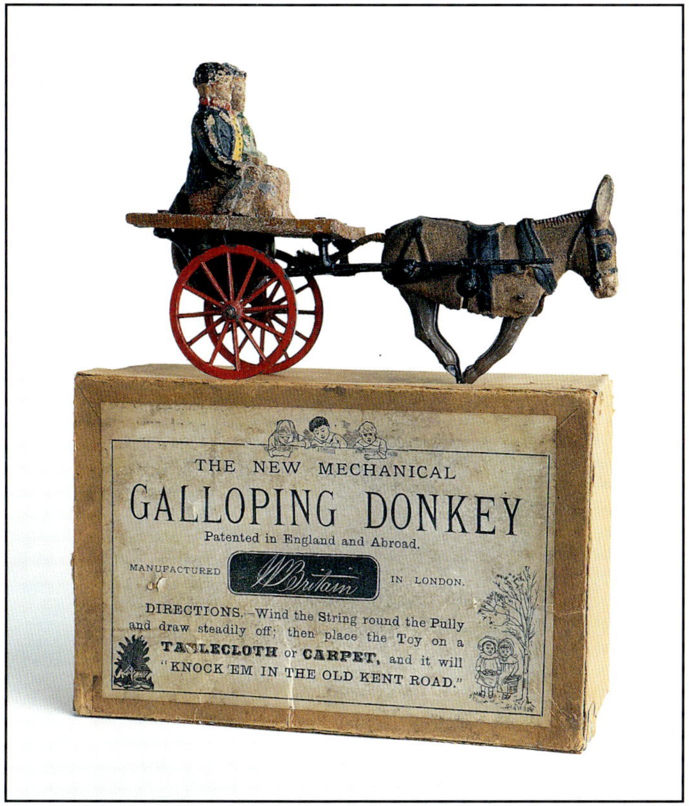

Galloping Donkey with cart driver and passenger in original box. A mechanical toy. *Courtesy Christie's Auction House, London/The Arnold Rolak Reference Collection.* $800-$900.

Chinaman with nodding head. A tinplate strip attached from inside the body to the neck enables the Chinaman to nod. Although no conclusive documentary evidence has been found, these items must be attributed to Britains with a possible German connection. *Courtesy Christie's Auction House, London/The Arnold Rolak Reference Collection.* $300-$400.

Chinaman and Monkey with nodding heads. The Monkey has the same body casting as the Chinaman. See previous illustration. *Courtesy Christie's Auction House, London/The Arnold Rolak Reference Collection.* $300-$400 and $200-$300.

Chinaman and Monkey with nodding heads, color variants. *Courtesy Christie's Auction House, London/The Arnold Rolak Reference Collection.* $300-$400 and $200-$300.

The Rotary Railway Express. *Courtesy Phillips Auction Company, Knowle England.* $1,500-$2,500.

The Steam Crane. *Courtesy Christie's Auction House, London/The Arnold Rolak Reference Collection.* $1,200-$1,400.

The Road Roller. Flywheel-driven tinplate and cast metal. Possibly manufactured as early as 1850. *Courtesy Christie's Auction House, London/The Arnold Rolak Reference Collection.* $500-$700.

Rotary Cyclist. *Courtesy New Cavendish Books/British Tin Toys. Marguerite Fawdry*. Unique.

The "Blondin" Cyclist. Blondin was a French tightrope walker and conquered Niagara Falls riding a penny farthing bicycle. *Courtesy Christie's Auction House, London/Britains Archive 1994.* $10,000-$12,000.

The Drinking Scotsman. This is the skeleton and shows part of the inner workings. Three Oriental heads, possibly for the tea drinking Mandarin. A larger head used either for the sailor money box or drinking Scotsman, and 3 other heads of uncertain origin. *Courtesy Christie's Auction House, London/Britains Archive 1994.* Unique.

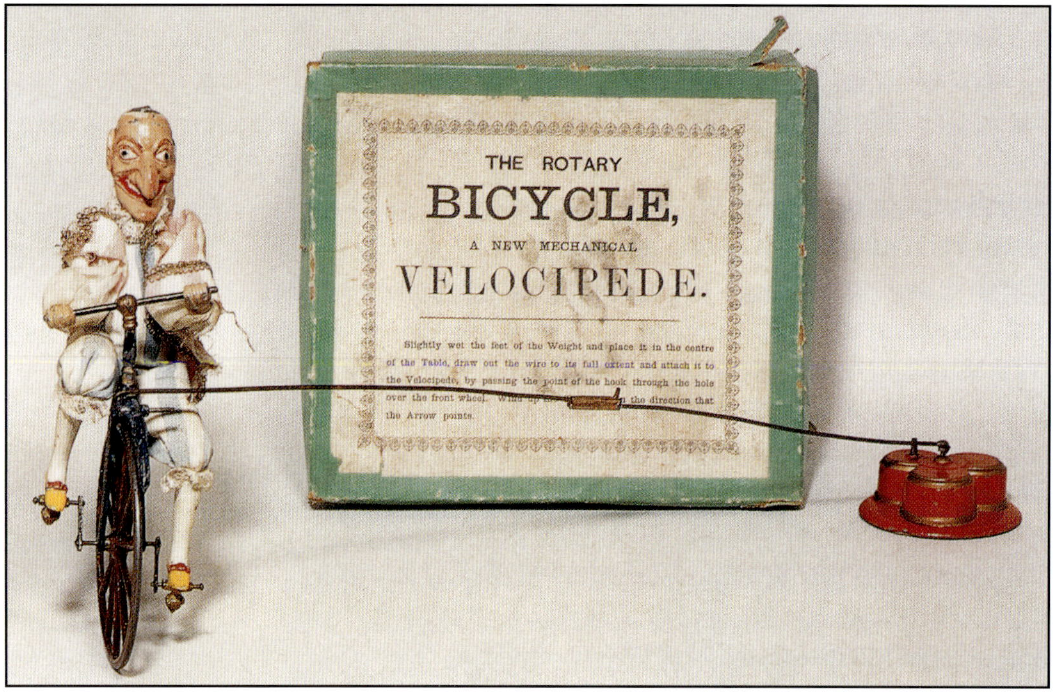

The Rotary Bicycle, Velocipede with Mr. Punch rider. *Courtesy New Cavendish Books/British Tin Toys. Marguerite Fawdry.* Unique.

The Drinking Scotsman. *Courtesy New Cavendish Books/David Pressland's Great Book of Tin Toys.* Unique.

Penny Cyclist, similar to the "Blondin" example. *Courtesy New Cavendish Books/ British Tin Toys. Marguerite Fawdry.* Unique.

Side view of Velocipede with Mr. Punch rider. *Courtesy New Cavendish Books/David Pressland's Great Book of Tin Toys.* Unique.

The Khedive. A clockwork item that smokes a hookah. *Courtesy New Cavendish Books/David Pressland's Great Book of Tin Toys.* Unique.

Other items shown in the 1880 Catalog (not illustrated): the Mechanical Dog; the Mechanical Bear; the Sailor Money Box; the Machine Gun; and the Indian Juggler.

3. Royalty, 1901

Britains used the Royal and Ceremonial occasions, Accession to the British Throne, Coronations, and Trooping-of-the-Color as an opportunity to issue souvenir and ceremonial items connected with the Royal Family. Starting as early as 1901 when a bust of King Edward the Seventh and Queen Alexandra was included in the souvenir section of the catalog, the tradition continued throughout the pre-War period. The 1937 Coronation proved to be the inspiration for a number of new items, including the State Coach. Post-War, the Coronation of Queen Elizabeth II and the Trooping-of-the-Color provided Britains with further new release opportunities.

Left: King Edward The Seventh. A rare large 3-5/8 inch gilt souvenir bust dated August 1901. *Courtesy Christie's Auction House, London/The Arnold Rolak Reference Collection.* $800-$900.

Right: King Edward the Seventh and Queen Alexandra standard-size souvenir busts dated August 1901. *Courtesy Christie's Auction House, London/The Arnold Rolak Reference Collection.* $100-$120

King Edward the Seventh and Queen Alexandra, standard-size Master Models/solid factory examples. *Courtesy Christie's Auction House, London/The Arnold Rolak Reference Collection.* Unique.

King Edward the Seventh in Field Marshall's uniform. Souvenir item in gilt #13, 45mm in size; an extremely rare item dated 1902. *Courtesy Christie's Auction House, London/The Arnold Rolak Reference Collection.* $600-$700.

Souvenir Catalog announcing new lines for the 1937 Coronation.

Set #1478 Miniature State Coach in gilt. Six inches long and produced at the time of the 1937 Coronation. Set #1503 was a smaller two horse version (not illustrated). Post-War, the correct number of eight horses was issued as catalog #44D. $60-$80.

A retail display card advertisement for miniature Coronation Crown stick pins #418B, 1937.

King George the Fifth silver finish souvenir bust and Queen Mary copper finish souvenir bust. *Courtesy Christie's Auction House, London/The Arnold Rolak Reference Collection.* $100-$120 each.

King George the Fifth silver finish souvenir bust #55D and Princess of Wales gilt finish souvenir bust #58D. The "D" Series figures were smaller than the standard size souvenir items. *Courtesy Christie's Auction House, London/The Arnold Rolak Reference Collection.* $100-$150 each.

Coronation Crown Stick Pins and a souvenir bottle of "English Coronation Anointing Oil" for export to the U.S.A., containing a Coronation crown without pin. *Courtesy Christie's Auction House, London/The Arnold Rolak Reference Collection.* $150 each.

Edward the Eighth. Upon the accession to the throne of Edward the Eighth, Britains started to adapt the mold for their Coronation Coach, which contained two figures (King and Queen). As Edward was not married, a single figure was produced in readiness for the 1937 issue Coronation Coach. This rare figure was a short lived issue as Edward was to abdicate and the mold was rendered obsolete. The eventual 1937 issue Coronation Coach contained figures of George the Sixth and Queen Elizabeth. Edward the Eighth was allocated Master Mold #970, while the dual casting of King and Queen seated was #539. *Courtesy Dr. Rob Wilson.* $120-$150.

Set #1473. King George the Sixth in Coronation Robes. Also issued in copper finish as set #1471 and gilt finish as #1472. The box shown had a paper sticker inside covering the inscription Edward the Eighth as Britains had printed this in readiness for the impending Coronation of Edward. Set #1506 Queen Elizabeth in coronation robes also issued in copper finish as set #1504 and gilt finish as set #1505. The gap in catalog sequence between the King (#1473) and Queen (#1506) was due to Britains having to hurriedly make the Queen figure when the unmarried Edward the Eighth did not accept the throne. *Courtesy Christie's Auction House, London/The Arnold Rolak Reference Collection.* $150-$180 each.

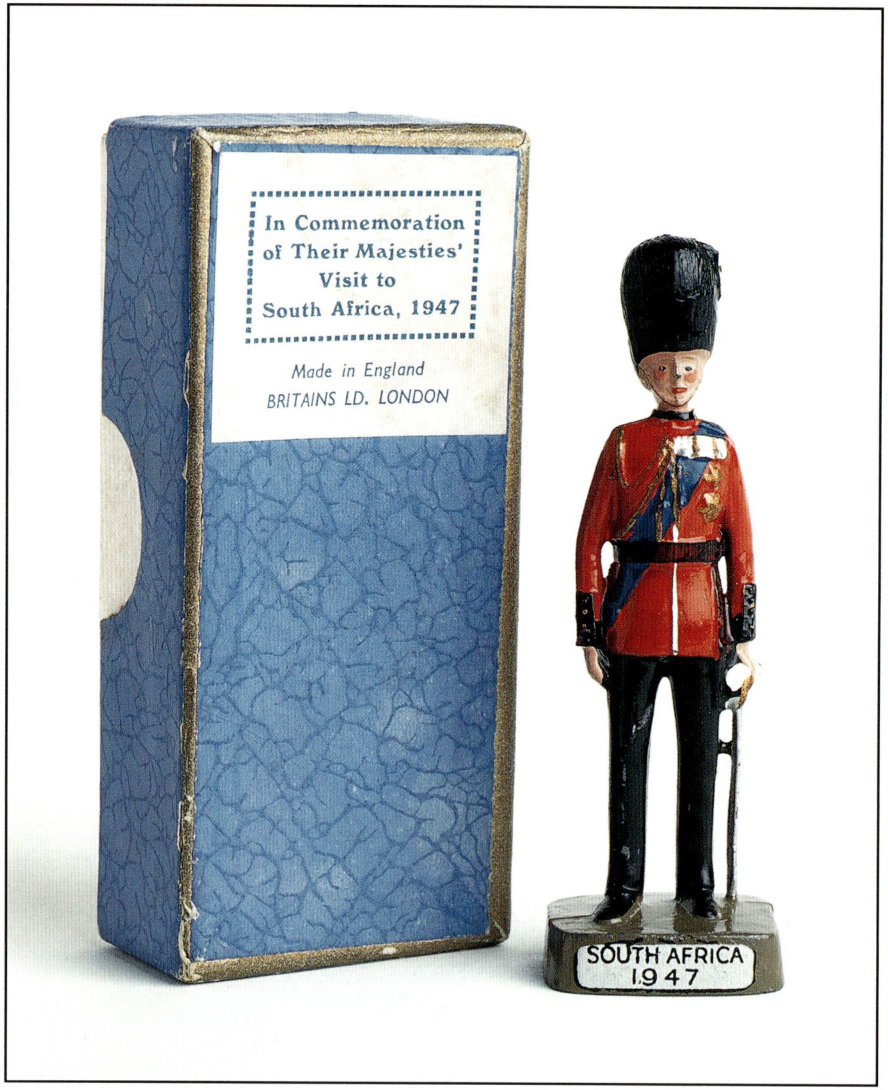

Set #1474 Coronation Chair of England as produced first for the 1937 Coronation in a box numbered 89D post-War coronation issue. *Courtesy Christie's Auction House, London/The Arnold Rolak Reference Collection.* $80-$100.

King George the Sixth in the uniform of Colonel in Chief of the Welsh Guards. Produced to commemorate the Royal visit to South Africa in 1947 with an inscription on the base to this effect. This figure was originally used as a Madame Tussaud issue representing H.R.H. Edward Prince of Wales as set #T6. See chapter on items made for other companies. *Courtesy Christie's Auction House, London/The Arnold Rolak Reference Collection.* $2,000-$2,400.

King George the Sixth in guardsman's uniform. A special paint issue with tan base indicating a possible 1937 Coronation connection. This item was not in general release. Adapted from the standard guards officer figure, it was allocated Master Model #1132B. *Courtesy Christie's Auction House, London/The Arnold Rolak Reference Collection.* $500-$600.

CORONATION OF

QUEEN ELIZABETH

1953

Coronation Brochure containing the plan for the 1953 Coronation procession.

Please address all Correspondence to the Company, and not to individuals. TELEPHONE LARKSWOOD 5881-2.

BRITAINS LIMITED

SUTHERLAND ROAD · WALTHAMSTOW

LONDON · E·17 · ENGLAND

METAL SOLDIERS AND ACCESSORIES

YOUR REF.

OUR REF.

MODEL FARMS—ZOO HUNTING—ETC.—ETC.

CORONATION OF HER MAJESTY

QUEEN ELIZABETH

We have pleasure in sending you a copy of our Coronation Lines Catalogue, which although not extensive in the number of lines it contains, is ample for Display and Souvenir purposes.

We particularly call your attention to the centre pages, where full information is given as to the order of the procession, as well as instructions for utilising our models to portray it.

The Catalogue itself has been produced in souvenir form and, if required, additional copies (for sale to the public) can be supplied at 50s. per 100.

BRITAINS LTD.

ASSOCIATE COMPANY W. HORTON (TOYS & GAMES) LTD.

Catalog insert with 1953 Coronation information.

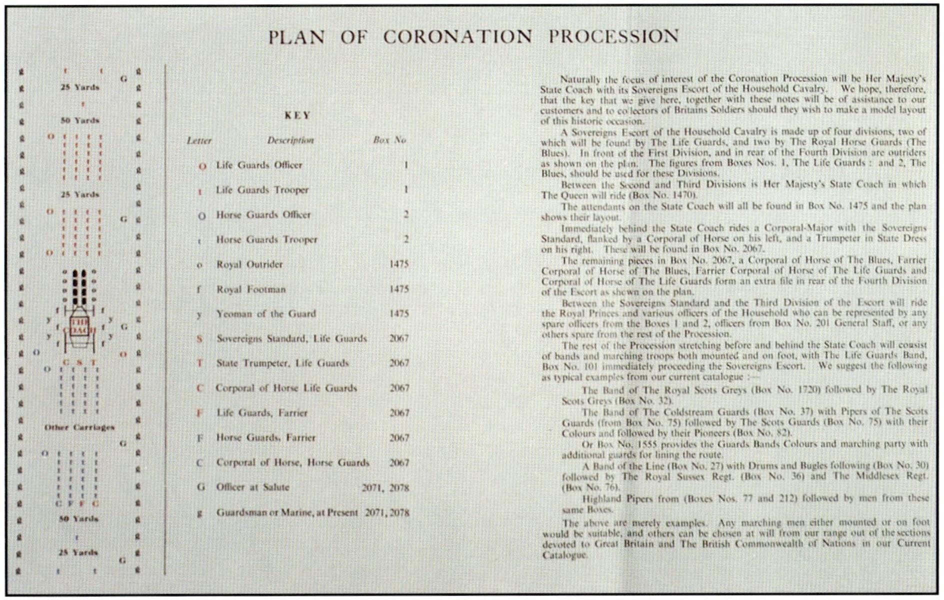

PLAN OF CORONATION PROCESSION

KEY

Letter	Description	Box No
O	Life Guards Officer	1
t	Life Guards Trooper	1
O	Horse Guards Officer	2
t	Horse Guards Trooper	2
o	Royal Outrider	1475
f	Royal Footman	1475
y	Yeoman of the Guard	1475
S	Sovereigns Standard, Life Guards	2067
T	State Trumpeter, Life Guards	2067
C	Corporal of Horse Life Guards	2067
F	Life Guards, Farrier	2067
F	Horse Guards, Farrier	2067
C	Corporal of Horse, Horse Guards	2067
G	Officer at Salute	2071, 2078
g	Guardsman or Marine, at Present	2071, 2078

Naturally the focus of interest of the Coronation Procession will be Her Majesty's State Coach with its Sovereigns Escort of the Household Cavalry. We hope, therefore, that the key that we give here, together with these notes will be of assistance to our customers and to collectors of Britains Soldiers should they wish to make a model layout of this historic occasion.

A Sovereigns Escort of the Household Cavalry is made up of four divisions, two of which will be found by The Life Guards, and two by The Royal Horse Guards (The Blues). In front of the First Division, and in rear of the Fourth Division are outriders as shown on the plan. The figures from Boxes Nos. 1, The Life Guards : and 2, The Blues, should be used for these Divisions.

Between the Second and Third Divisions is Her Majesty's State Coach in which The Queen will ride (Box No. 1470).

The attendants on the State Coach will all be found in Box No. 1475 and the plan shows their layout.

Immediately behind the State Coach rides a Corporal-Major with the Sovereigns Standard, flanked by a Corporal of Horse on his left, and a Trumpeter in State Dress on his right. These will be found in Box No. 2067.

The remaining pieces in Box No. 2067, a Corporal of Horse of The Blues, Farrier Corporal of Horse of The Blues, Farrier Corporal of Horse of The Life Guards and Corporal of Horse of The Life Guards form an extra file in rear of the Fourth Division of the Escort as shown on the plan.

Between the Sovereigns Standard and the Third Division of the Escort will ride the Royal Princes and various officers of the Household who can be represented by any spare officers from the Boxes 1 and 2, officers from Box No. 201 General Staff, or any others spare from the rest of the Procession.

The rest of the Procession stretching before and behind the State Coach will cousist of bands and marching troops both mounted and on foot, with The Life Guards Band, Box No. 101 immediately proceeding the Sovereigns Escort. We suggest the following as typical examples from our current catalogue :—

The Band of The Royal Scots Greys (Box No. 1720) followed by The Royal Scots Greys (Box No. 32).

The Band of The Coldstream Guards (Box No. 37) with Pipers of The Scots Guards (from Box No. 75) followed by The Scots Guards (Box No. 75) with their Colours and followed by their Pioneers (Box No. 82).

Or Box No. 1555 provides the Guards Bands Colours and marching party with additional guards for lining the route.

A Band of the Line (Box No. 27) with Drums and Bugles following (Box No. 30) followed by The Royal Sussex Regt. (Box No. 36) and The Middlesex Regt. (Box No. 76).

Highland Pipers from (Boxes Nos. 77 and 212) followed by men from these same Boxes.

The above are merely examples. Any marching men either mounted or on foot would be suitable, and others can be chosen at will from our range out of the sections devoted to Great Britain and The British Commonwealth of Nations in our Current Catalogue.

1953 Coronation procession plan.

Set #1470. Her Majesty's State Coach with team of eight Windsor grey horses and postillions in state livery. Master Model Number Postillion and Horse #1129B, Trace Horse #1130B, King and Queen seated #1131B. *Courtesy Christie's Auction House, London.* $400-$450.

Set #2094. State open road landau with team of six Windsor grey horses and H.M. Queen Elizabeth and the Duke of Edinburgh. Master Model Numbers: H.M. the Queen #1376B; H.R.H. The Duke of Edinburgh #1377B; Footmen #1378B; Near side Landau horse #1380B; Off-side Landau horse #1381B; Landau #1382B. *Courtesy Christie's Auction House, London.* $450-$500.

Coronation procession with State attendants from set #1475. Master Model Numbers: Footman #419B Outrider/Groom #420B, and Yeoman #728B. *Courtesy Christie's Auction House, London/The Arnold Rolak Reference Collection.* $35-$45 each.

A rare brown painted version of the State Landau horse. $80-$100.

Set #2065. H.M. Queen Elizabeth, Colonel in Chief of the Grenadier Guards. Trooping-of-the-Color. *Courtesy Christie's Auction House, London/The Arnold Rolak Reference Collection.* $150-$200.

Garter King of Arms mounted an uncataloged special paint finish item reputed to have been commissioned by a London department store in 1953. *Courtesy Christie's Auction House, London/The Arnold Rolak Reference Collection.* $500-$600.

Set #2065. H.M. Queen Elizabeth, Colonel in Chief of the Grenadier Guards. Trooping-of-the-Color riding a white horse. The Queen rode this horse only once and the set was intended for the Canadian market, hence its rarity. *Courtesy Christie's Auction House, London/The Arnold Rolak Reference Collection.* $250-$350.

Set #2065 H.R.H. Princess Elizabeth, Colonel of The Grenadier Guards, Trooping-of-the-Color issued before the Coronation. This rare box gives the future Queen the title of Princess. Master Mold #993 was allocated to this item to distinguish it from set #2065 H.M. the Queen. *Courtesy Christie's Auction House, London/The Arnold Rolak Reference Collection.* $350-$400.

4. Novelties/Souvenirs, 1901

Miniature toy soldiers one sixth of the size of the standard toy soldier, 54mm, made in 1922 for Queen Mary's doll house, Windsor. *Courtesy Christie's Auction House, London/Britains Archive 1994.* Unique.

1924 saw the inclusion of a list of household utensils and, although novelty items by way of soldiers, royal busts, and the Football Association Cup had been advertised in 1901 and 1915, the novelty range was not promoted in full. In 1922, tiny soldiers, one sixth the size of a 54mm figure, were produced for Queen Mary's doll house. From 1924 onwards, a variety of novelty items were produced, and in the case of the household utensils, carried into Post-War production.

Football Association Cup. *Courtesy Philip Dean.* $1,200-$1,500.

Novelty jointed footballer pin/brooch, Master Mold #1291, with Association Football Cup. *Courtesy Christie's Auction House, London/Britains Archive 1994.* Pin $400-$450.

Household utensils, pre-War products: Galvanized pail #1U, Black coal scuttle with tongs #2U, Black griller #8U, Water jug #10U, Washtub #11U. *Courtesy Christie's Auction House, London/The Arnold Rolak Reference Collection.* $30-$40 each.

Coster Barrow #32/33. Also comes painted gilt, yellow, or blue. *Courtesy Christie's Auction House, London/The Arnold Rolak Reference Collection.* $50-$60.

Household Furniture: Baby's highchair #26, Rocking chair #25, Push chair #27. *Courtesy Christie's Auction House, London/The Arnold Rolak Reference Collection.* $60-$80 each.

Step ladders #31. Also came painted gilt. $35-$45 each.

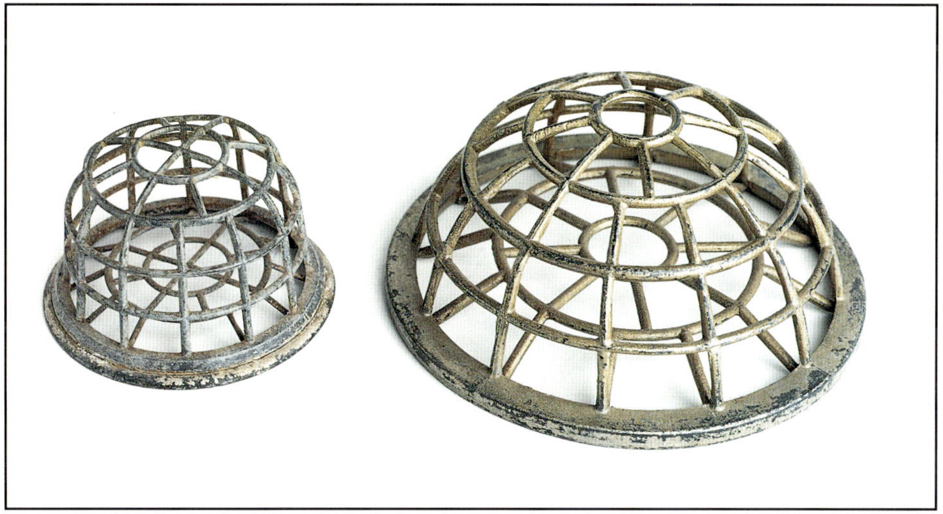

Ideal flower supports, small #1526, large #426. Used for flower arranging. *Courtesy Christie's Auction House, London/The Arnold Rolak Reference Collection.* $100-$120 each.

Miniature Kitchen Utensils set #1. *Courtesy Christie's Auction House, London/The Arnold Rolak Reference Collection.* $220-$250.

Miniature Kitchen Utensils set #2. *Courtesy Christie's Auction House, London/The Arnold Rolak Reference Collection.* $220-$250.

Miniature kitchen utensils with original box. Gilt teapot #9, Black saucepan #21, Copper kettle #22, Black boiler #30, Steps #31, Coffeepot, Watering can, Copper Boiler #29, Black stew pan #30. *Courtesy Ray Haradin.* $250-$300.

Locomotive paperweights. Canadian National Railways No. 6000, London and North Eastern Railway Flying Scotsman, as souvenir of the motion picture "Flying Scotsman," London and North Eastern Railway High Pressure Compound #10000 in rare box, and a LMS Streamline Engine named "Coronation". *Courtesy Christie's Auction House, London/The Arnold Rolak Reference Collection.* $100-$120 each.

Pots and Pans set #1HU, post-War issue. *Courtesy Christie's Auction House, London/The Arnold Rolak Reference Collection.* $300-$350.

Morphy Richard Irons. Right: #HU54 Morphy Richard's auto control safety iron with original box. Left: same item re-boxed without the Morphy Richard's brand name. These items were the first advertising items for other companies produced post-War. *Courtesy Christie's Auction House, London/The Arnold Rolak Reference Collection.* $80-$90 each.

5. Footballers, 1904

Footballers were first introduced into the Britains catalog in 1904 and comprised of twelve teams. Corinthians, an amateur team, were discontinued prior to 1922 when sixteen additional teams were added. Plymouth Argyle was added in 1936 to bring the compliment of teams up to twenty-eight.

Without the assistance of finding complete tied-in-the-box sets, identification of team colors can be somewhat difficult. Many teams played in similar colored strips and sometimes changed strips if the opposing team's strip was similar. This made things much easier for the referee in an actual football match to distinguish between teams. Britains also undertook commissions from individuals to paint teams in colors other than those in the standard range of twenty-eight.

Football figures were also exported to Australia and sold via Walther and Stevenson's Department Store in Sydney. Once in Australia, the figures were issued and sold as miniature footballers in Sydney rugby league colors. These colors do not resemble any of the British team colors and, therefore, must also have been specially commissioned.

The identification of teams in the following illustrations is as accurate as the information to hand will allow.

The first twelve teams issued in 1904 were Aston Villa, Blackburn Rovers, Corinthians, Derby County, Everton, Manchester City, Notts County, Preston North End, Sheffield Wednesday, Sunderland, Tottenham Hotspur, and Woolwich Arsenal (later known only as Arsenal).

The early twenties saw the addition of sixteen teams and the discontinuation of Corinthians. The new sixteen were: Birmingham City, Bolton Wanderers, Bradford City, Burnley, Bury, Cardiff City (the only Welsh team playing in the football league), Chelsea, Huddersfield Town, Liverpool, Manchester United, Middlesborough, Oldham Athletic, Stoke City, West Bromwich Albion, and Wolverhampton Wanderers. With the further addition of Plymouth Argyle in 1936, twenty-eight teams remained in the catalog until 1941. The Football Series was not revived post-War. Master Mold numbers for each basic casting were #479 Referee, #480 Linesman, #481 Goalkeeper, #482 Fullback, #483 Halfback, #484 Forward, #553 Corner Flag, #554 Goal Post, and #556 Football.

Examples of all colors for the teams in the standard catalog range are shown in the following illustrations with the exception of: #174B red and white vertical stripes and blue shorts, unknown team; #178B black and white stripes and white shorts, unknown team; #180B Bolton Wanderers; #183B Liverpool or Manchester United; #190B Middlesborough; and #211B Dundee, Scottish Club.

Joe Wallis, who has carried out much research into Britains figures, had an excellent article published in *Old Toy Soldier Newsletter* magazine in 1989, *Volume 13, #8*. Joe is the author of two toy figure books *Armies of the World* and *Regiments of All Nations*.

This chapter was compiled with the assistance of Alf Mussett and Dick Charlesworth, both of whom have collected football figures for many years. I appreciate their assistance and I am grateful to them for sharing their knowledge.

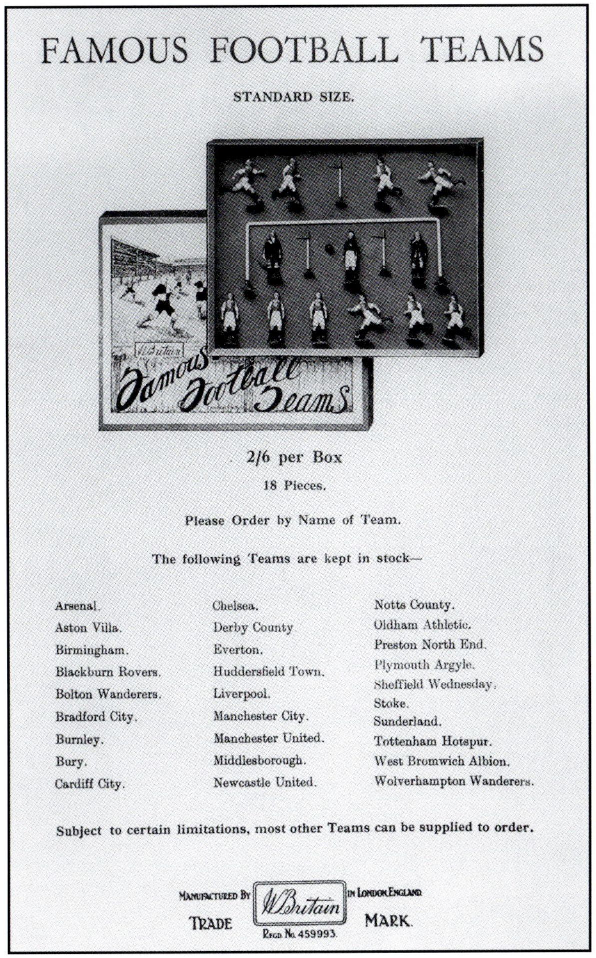

FAMOUS FOOTBALL TEAMS

STANDARD SIZE.

2/6 per Box

18 Pieces.

Please Order by Name of Team.

The following Teams are kept in stock—

Arsenal.	Chelsea.	Notts County.
Aston Villa.	Derby County	Oldham Athletic.
Birmingham.	Everton.	Preston North End.
Blackburn Rovers.	Huddersfield Town.	Plymouth Argyle.
Bolton Wanderers.	Liverpool.	Sheffield Wednesday.
		Stoke.
Bradford City.	Manchester City.	Sunderland.
Burnley.	Manchester United.	Tottenham Hotspur.
Bury.	Middlesborough.	West Bromwich Albion.
Cardiff City.	Newcastle United.	Wolverhampton Wanderers.

Subject to certain limitations, most other Teams can be supplied to order.

Manufactured By W. Britain in London England
TRADE MARK
Regd No. 459993.

The "Right Angle" Football Game

JUST THE GAME FOR THE WINTER EVENINGS

Skill, Sport and variety in every Game.

No. 1528 An Interesting and Exciting Game for two or more players, based on the regulation game. It consists of a Football Field with twelve Men (six per side), Football and two Dice, and full instructions for the Game. 2/6

Brightly boxed and labelled. Board open measures 24 × 14 ins.

Manufactured By W. Britain in London England
TRADE MARK
Regd No. 459993.

Left: Famous Football Teams. Second version illustrated box and team. This catalog page dates from around 1938 and lists the teams available, as well as announcing the availability of custom team products.

Right: The "Right Angle" Football Game Catalog #1528. A reduced price version of Catalog #1524. Contains board marked out in squares, dice football, and twelve players. $1,500-$2,000.

Famous Football Teams. Very rare, possibly unique, Clapton Orient. A special order paint set. Clapton Orient was the local team closest to the Britains Factory, later named Leyton Orient, and then simply Orient. This set was formerly part of the Alf Mussett Collection and now resides in the Philip Dean Reference Collection. $1,500-$1,800.

Famous Football Teams, Arsenal. $600-$800.

Famous Football Teams, Chelsea. $600-$800.

Famous Football Teams, Aston Villa. $600-$800.

Famous Football Teams, Newcastle United. $600-$800.

Famous Football Teams, Blackburn Rovers. $600-$800.

Famous Football Teams, West Bromich Albion. $600-$800.

Goal Post Master Model #256B, Corner Flag #257B, Football #258B, Referee #254B, Linesman #255B. Goal Post, $80-$100; Corner Flag, $55-$65; Football, $60-$80; Referee and Lineman, $100-$120 each.

Goalkeepers Master Model #189B. The figure in the center is a Sunderland goalkeeper with jersey painted in early first version style, the same as the team colors. Second version goalkeepers, left and right, wore green or blue jerseys. $85-$90 each.

Linesman Master Model #255B, color variants. $100-$120 each.

Fullbacks. Walking Footballers Master Mold #482 was used for this casting. Left to right: West Bromich Albion, Master Model #175B; Derby County, Master Model #181B; Sheffield Wednesday, Master Model #176B; Birmingham City, Master Model #210B; Aston Villa, Master Model #187B; Newcastle United, Master Model #177B; and Blackburn Rovers, Master Model #179B. $60-$65 each.

Fullbacks. Walking Footballers, Master Mold #482. Left to right: Arsenal, Master Model #191B; Clapton Orient, special order paint; Accrington Stanley, special order paint; Sunderland, Master Model #172B; Stoke City, Master Model #173B; unknown team; and Manchester City, Master Model #185B. $60-$65 each. Clapton, $100-$150.

Halfbacks. Running Footballers, Master Mold #483. Left to right: Sunderland, Master Model #172B; unknown team; Manchester City, Master Model #185B; Aston Villa, Master Model #187B; and two unknown, possibly European, teams. $60-$65 each.

Above: Fullbacks. Walking Footballers, Master Mold #482. Left to right: Plymouth Argyle, Master Model #270B; Bradford City, Master Model #188B; Blackpool, Master Model #184B; Everton, Master Model #182B; unknown team; Wolverhampton Wanderers, Master Model #186B; and an unknown team. $60-$65 each.
Below: Halfbacks. Running Footballers, Master Mold #483 allocated to this casing. Left to right: West Bromich Albion, Master Model #175B, Birmingham City, Master Model #210B; Aston Villa, Master Model #187B; and Newcastle United, Master Model #177B. $60-$65 each.

Above: Forward. Running Footballers kicking, Master Mold #484 allocated to this casting. Left to right: Blackburn Rovers, Master Model #179B; Arsenal, Master Model #191B; Clapton Orient and Accrington Stanley special order teams; Birmingham City, Master Model #210B; and Sunderland, Master Model #172B. $60-$65 each. Clapton, $100-$150.
Below: Forward. Running Footballer kicking, Master Mold #484. Left to right: Manchester City, Master Model #185B; unknown team; Bradford City, Master Model #188B; Wolverhampton Wanderers, Master Model #186B; and two unknown, possibly European, teams. $60-$65 each.

Clapton Orient Fullback, Master Mold #482; Halfback, Master Mold #483; Forward, Master Mold #484. *Courtesy Philip Dean Reference Collection.* $100-$150 each.

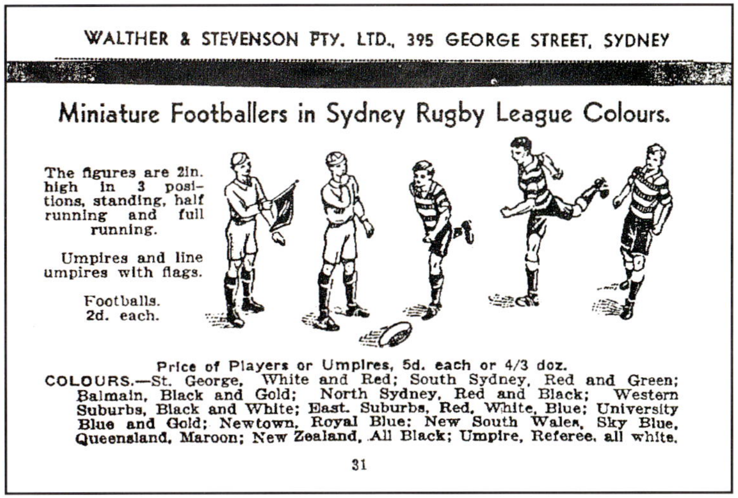

WALTHER & STEVENSON PTY. LTD., 395 GEORGE STREET, SYDNEY

Miniature Footballers in Sydney Rugby League Colours.

The figures are 2in. high in 3 positions, standing, half running and full running.

Umpires and line umpires with flags.

Footballs. 2d. each.

Price of Players or Umpires, 5d. each or 4/3 doz.

COLOURS.—St. George, White and Red; South Sydney, Red and Green; Balmain, Black and Gold; North Sydney, Red and Black; Western Suburbs, Black and White; East. Suburbs, Red, White, Blue; University Blue and Gold; Newtown, Royal Blue; New South Wales, Sky Blue, Queensland, Maroon; New Zealand, All Black; Umpire, Referee, all white.

31

Catalog page from Walther and Stevenson's Department Store, Australia, showing line drawings of footballers destined to be sold as Australian rugby figures. *Courtesy Robin Yates.*

Base of football figure with paper sticker dated 1904. Paper stickers were added as copyright protection, later the wording was embossed on each base. *Courtesy Philip Dean.*

Factory sample cards showing Footballers Master Model numbers. The two cards on the left are examples of Dutch football team colors and have "V" prefix reference numbers. The card top right shows three rare Glasgow Celtic figures, Master Mold #209, along with English football players already shown in previous illustrations. *Courtesy Christie's Auction House, London/Britains Archive.* Unique.

Factory sample card for Australian football/rugby items. The referee in all white, bottom right, was allocated Master Model #1520B to distinguish him from the standard issue referee #254B. The absence of a goalkeeper confirms that these items were to be sold as rugby players. Unique.

Football Association Cup, Master Mold #596; Lid, Master Mold #597. This item was issued in the Novelty Range during the early 1900s. *Courtesy Christie's Auction House, London/The Arnold Rolak Reference Collection.* $1,200-$1,500.

Football Association Cup. Close up of lid with football on figure. *Courtesy Philip Dean Reference Collection.*

6. Salvation Army, 1906

Gilt bust of General William Booth, founder of the Salvation Army, dated March 20, 1906. A silver version is also known to exist along with a matte finish example in rust red. *Courtesy Christie's Auction House, London/The Arnold Rolak Reference Collection.* $800-$900.

Salvation Army figures set #10 were available for a short period, 1906-1908, and included Officers, Color Bearer and Buglers. Female Salvation Army figures in long dresses were available in the same period via set #14.

Second version Salvation Army were released in 1934. The female figures dress was shortened to keep in line with the fashion of the day. Full bands in both red and blue uniforms were included in the range which did not appear in any Britains catalog. The figures were available direct from Salvation Army Headquarters in Judd Street, London.

Rare Salvation Army set #10 issued in 1906 with oval-based officers, buglers, and standard bearers. *Courtesy Christie's Auction House, London/The Arnold Rolak Reference Collection.* $4,000-$4,500.

Rare boxed Salvation Army set #10. *Courtesy Hank and Doris Sigourney*. $6,000-$7,000.

Salvation Army buglers close-up from set #10. *Courtesy Christie's Auction House, London/The Arnold Rolak Reference Collection*. $300-$400 each.

Salvation Army standard bearers and officers. Left to right: Oval-based standard bearer from set #10; standard bearer in red tunic, Master Model #202; standard bearer in blue tunic, Master Model #203; square and round based officers, Master Model #204B. A Salvation Army private (not illustrated) was allocated Master Model #205B. Oval-based figures date from 1906 to 1908, square based figures are from 1934 to 1941. *Courtesy Philip Dean Reference Collection*. Oval based $400-$500. Others $200-$250 each.

Salvation Army women from set #14. Left to right: First long dress version and second shorter dress version marching empty handed Master Model #206B. First long dress version and second shorter dress version carrying war cry Master Model #207B. First long dress version and second shorter dress version tambourine player Master Model #208B. *Courtesy Philip Dean Reference Collection*. Long Dress Variant, $250-$350. Short Dress Variant, $180-$250 each.

Salvation Army Band, rare set #1317. 24-piece band led by Bandmaster in blue tunic with cornet enhanced by a standard bearer. *Courtesy Philip Dean Reference Collection*. $4,500-$5,000.

Rare Salvation Army Bandsmen in blue uniforms. Left to right: Bandmaster, #227B; Cornet, Master Model #228B; Euphonium, #229B; Double Bass, #230B; Trombone, #231B; Cymbals, #232B; Tenor Horn, #233B; Monstre Bass, #234B; Side Drum, #235B; Big Drum, #236B. *Courtesy Philip Dean Reference Collection*. $240-$260 each.

#1316. Very rare Salvation Army set in blue uniforms, comprising a full band and personnel. *Courtesy Philip Dean Reference Collection*. $7,000-$8,000.

Top right: Salvation Army Bandsmen. Left to right: Cornet, Master Model #196B; Euphonium, #197B; Bass, #198B; Trombone, #199B. *Courtesy Philip Dean Reference Collection*. $120-$150 each.
Center right: Salvation Army Bandsmen. Left to right: Side Drummer, Master Model #200B; Bass Drum, #201B; Cymbals, #224B. *Courtesy Philip Dean Reference Collection*. $120-$150 each.
Bottom right: Salvation Army Bandsmen. Left to right: Tenor Horn, Master Model #225B; Monstre Bass, #226B; Bandmaster, Cornet blue uniform, #227B. *Courtesy Philip Dean Reference Collection*. $120-$150 each.

7. Civilians, 1908

Figures from set #168 Civilians/Pedestrians showing all the variations used in the set between 1908 and 1941. The Edwardian Lady was replaced by the figure, second left, nicknamed "Flapper". This item was included in the Railway Series as #806 where it also replaced the old #806 Edwardian Lady and was employed also in the Farm Range as Young Lady #559. See Farm and Railway chapters for examples. The standing man with pipe and standing man with newspaper also had a dual, interchangeable role, appearing from time to time in this set and as railway passengers. The Policeman was issued as #808 in the Railway Series and post-War as #776 in the Farm Range. In this set, his armband is red and white indicating City of London Police. The Yachtsman was #817 in the Railway Series. The other figures are shown in the following illustration. *Courtesy Philip Dean Reference Collection*. Prices: see individual figures in this chapter and Railway chapter.

1908 saw the introduction of set #168 Civilians. This title was later changed to Pedestrians. All were new figures and, with the exception of the items described in the first illustration, were employed within the Railway Series from 1909 onwards.

Civilians set #168. Issued first in 1908, the title of the set was changed to Pedestrians in 1936 when it was included with the Traffic Series. Man with Panama hat, left, man walking with pipe, four from the left, and chauffeur, far right, were available only in this set. The others were also employed as railway passengers (see Railway chapter.) *Courtesy Christie's Auction House, London/The Arnold Rolak Reference Collection.* $1,500-$1,800.

Man walking with pipe allocated Master Model #889B, available only in set #168 and not given a catalog number. *Courtesy Philip Dean Reference Collection.* $80-$90.

Top right: Edwardian Woman, Catalog #806B, color variations. See also Railway chapter. $85-$90.

Bottom right: Man in Panama hat allocated Master Model #887B, available only in set #168 and not given a catalog number. *Courtesy Philip Dean Reference Collection.* $90-$120.

Chauffeur with lamp allocated Master Model #890B, only available in set #168 and not given a catalog number. Two other versions of this figure exist. An oval based version (see *The Great Book of Hollow Cast Figures* by Norman Joplin, Published by New Cavendish Books) and an adaptation of the Tall Zoo Keeper figure in a green uniform, but without a lamp. (Neither illustrated.) *Courtesy Philip Dean Reference Collection.* $90-$120.

8. Railway, 1909

Railway Staff and Passengers in standard 54mm size were first introduced in 1909 with two sets, Catalog numbers 155 and 158. "O" Gauge (45mm) figures appeared first in 1933. All were discontinued in 1941. Passengers from set #158 were also issued in Civilian/Pedestrian set #168. See chapter covering police and road vehicles. A newly designed set of Railway Staff and Passengers was issued in 1954 and was deleted after 1959 (see set #1R).

Complete Railway Station set #158. *Courtesy Ray Haradin*. $2,400-$2,600.

Set #158 box artwork for Railway Station set. *Courtesy Dr. Rob Wilson*.

Railway Porter to push trolley Catalog #800, Railway Porter with moveable arm to carry luggage Catalog #801. Left to right: with trolley, $60-$70; without trolley, $45-$50; with luggage, $75-$80; without luggage, $45-$50.

Railway Station Staff set #155. Displayed out of box. *Courtesy Philip Dean Reference Collection*. See previous illustration.

Station Master, Catalog #802, Ticket Collector, Catalog #805. This is the same figure with a plain arm replacing the arm holding document from the Station Master to the Ticket Collector. The Station Master is also identified by his grey beard as opposed to the Ticket Collector's brown beard. $50-$55 each.

Railway Station Staff set #155. *Courtesy Ray Haradin*. $700-$900.

Top left: Guard with flag, Catalog #803, flanked by two casting versions of Guard with lamp, Catalog #804. The same figure was used for both with an arm change. $55-$60 each.

Top right: Young Lady Passenger with handbag, Catalog #806, sometimes nicknamed "Flapper". This figure also appeared in set #168, Civilians/ Pedestrians, replacing the Victorian Lady figure, and in the Farm Range as Catalog #559; a version in white representing a Tennis Player was also available. See also Civilian and Farm chapters for further examples. Master Mold #159 was allocated to this item. $25-$30 each. Tennis variant, $80-$90.

Bottom left: Edwardian Lady Passengers, Catalog #806. This figure was also included in early issues of set #168 Civilians/Pedestrians and was later replaced in both the Railway and Civilian Series by the Young Lady Passenger figure shown in the following illustration. See Civilian chapter for more examples and color variants. $85-$90 each.

Bottom right: Man standing with pipe, Catalog #807. This Railway Passenger also appeared in set #168, Civilian/ Pedestrians, and used the same casting as Man standing with newspaper as shown in the following illustration. See Civilian chapter. $80-$85 each.

Left: Man standing with folded newspaper allocated the same Catalog #807 as Man standing with pipe. Both use the same casting with only an arm change. This item also appeared in set #168 Civilians/ Pedestrians. See Civilian Chapter. $80-$85 each.

Engine Driver, Catalog #809, and Stoker with Shovel, Catalog #810. The same casting was used for both items with only a slight variation in the Stoker's arms. A Cricketer was adapted from this casting also. See relevant chapter for further details. $65-$70 each.

Below: Porters Trolley, Catalog #811; Trunk, Catalog #812; Dress Basket, Catalog #813; Portmanteaux, Catalog #814; Golf Sticks, Catalog #815; and Rug and Sticks, Catalog #816. Trolley $30-$35; Trunk $30-$35; Dress Basket $30-$35; Portmanteaux $30-$35; Golf Sticks $50-$55; Rug and Sticks $45-$50.

Left: Policeman standing wearing helmet, Catalog #808. This item was also included in set #168 Civilians/ Pedestrians and as Catalog #776 in the post-War Farm Range. The armband on the sleeve is usually blue and white. However, a rarer red and white armband denoting City of London Police is sometimes found in set #168. See Civilian and Farm chapters for further examples. $30-$35. Red and white armband variant, $70-$80.

Yachtsman with pipe and coat over arm, Catalog #817. Also issued in set #168 Civilians/Pedestrians. See Civilian chapter for further details. $85-$90.

Lady Railway Passengers with stick and umbrella issued post-War only and allocated Master Mold #1164. $150 each.

Man with pipe issued post-War only, allocated Master Mold #1165. $150.

Model Railway Series set #1R issued post-War. This set contained adaptations of pre-War figures along with a number of new figures. The new figures were a Lady Passenger with stick or umbrella in moveable arm allocated Master Mold #1164. A Man standing with pipe without hat, Master Mold #1165. A Military Policeman allocated Master Model #1332B and adapted from the Military issue Home Guard figure out of set #1918. The other adapted items included a new large peak cap for the porters. A ticket collector and guard with green flag adapted with arm changes and paint variations from the Short Zoo Keeper figure and a Station Master with arm and paint adaptation from the Tall Zoo Keeper. *Courtesy Philip Dean Reference Collection.* $2,500-$2,800.

#1R. Porter carrying luggage adapted from the pre-War Porter #801. This example has a larger head with 1950s type cap. *Courtesy Dr. Rob Wilson*. $80-$100.

#1R. Porter pushing trolley adapted from the pre-War Porter #800 with new style cap and larger head. *Courtesy Dr. Rob Wilson*. $80-$100.

#1R. Ticket Collector. An adaptation and paint variant of the small Zoo Keeper. *Courtesy Dr. Rob Wilson*. $80-$100.

#1R. Guard with red flag. An adaptation and paint variant of the small Zoo Keeper with an arm change. The #1R set contained this figure with a green flag. This example is rare, as it has a red flag. *Courtesy Dr. Rob Wilson*. $100-$120.

#1R. Station Master. An adaptation and paint variant of the large Zoo Keeper. *Courtesy Dr. Rob Wilson*. $80-$100.

"O" Gauge Railway Series Golfer Catalog #819 and included in the larger set #1256 (not illustrated). *Courtesy Brian Holmes*. $80-$100.

"O" Gauge Railway set #1423. Designed to fit the scale of Hornby Trains, 42mm in height. Left to right, Plate Layer, Catalog #818, Ticket Collector, Catalog #822, Porter for Trolley, Catalog #821, Guard with Flag, Catalog #820, Lady Passenger, Catalog #823 and Policeman in Helmet, Catalog #824. The baggage and porters trolley were allocated the same Catalog #825. *Courtesy Philip Dean Reference Collection*. $500-$600.

9. Indians, 1908 and Cowboys, 1913

North American Indians, as Britains first described them, were originally issued in 1908 with the introduction of sets #150 and #152. Cowboys did not follow until 1913, the first set being #179. American Cowboys was the title used in these earlier years with the abbreviated and more widely used Cowboys and Indians term becoming more generally used to describe these ranges. Combination sets of Cowboys and Indians were normally given the title and referred to as Wild West. Many of the standard 54mm Cowboy and Indian figures remained until the end of lead production in 1966 with a number being employed in the Picture Pack Range between 1954 to 1959.

As well as being issued in the many different boxed sets, some Cowboys and Indians were also released as single items in the Picture Pack range and allocated Master Model "B" suffix.

Summary of Boxed Sets Entitled Cowboys, Indians or Wild West

One or more of all of the first grade paint single figures of Cowboys and Indians would have been included in at least one of the following boxed sets.

Pre-War Only Issues: *indicates not illustrated
- #184 Cowboys mounted and foot*
- #185 Wild West display*
- #210 Indians mounted and foot and trees*
- #244 Indians mounted and foot*
- #245 Cowboys mounted and foot
- #256 Cowboys mounted and foot*
- #257 Indians mounted and foot*
- #273 Cowboys and Indians mounted and foot
- #274 Indians mounted and foot*
- #275 Cowboys mounted and foot*
- #277 Indians mounted and foot*
- #278 Cowboys mounted and foot*
- #281 Indians mounted and foot with tent*
- #282 Cowboys mounted and foot with tent*
- #297 Indians mounted and foot*
- #298 Cowboys mounted and foot*
- #305 Indians mounted and foot*
- #306 Cowboys mounted and foot*
- #367 Uncataloged mounted Indians (6)
- #1252 Cowboys on foot*
- #1311 Cowboys and Indians on foot*
- #1312 Cowboys and Indians mounted*
- #1508 Texas Rangers. Standard issue mounted Cowboys (5 figures) given Texas Rangers title
- #1509 Texas Rangers. Standard issue mounted Cowboys given the Texas Rangers title (8 figures)

Pre- and Post-War Sets: *indicates not illustrated
- #150 (#9189) Indians on foot, pre-War*
- #152 (#9289) Indians mounted, pre-War*
- #179 (#9238) Cowboys mounted
- #183 (#9188) Cowboys on foot
- #208 (#9389) Indians mounted and foot, pre-War*
- #209 (#9388) Cowboys mounted and foot, pre-War*
- #272 (#9390) Cowboys and Indians mounted and foot, pre-War*

Post-War Sets: *indicates not illustrated
- #2034 Covered wagon "Prairie Schooner" with Pioneer and Wife
- #2042 Covered wagon set, "Prairie Schooner" with Cowboy Escort and Indian.
- #2043 Rodeo set
- #2061 Wild West display
- #2145 Cowboys mounted and foot*
- #2146 Indians mounted and foot.
- #2163Q Mazak uncataloged set for Australian market

9000 Numbered Sets Showing Original Number. All in cellophane fronted boxes
- #9188 (#183) Cowboys on foot
- #9189 (#150) Indians on foot
- #9288 (#179) Cowboys mounted
- #9289 (#152) Indians mounted
- #9388 (#209) Cowboys mounted and foot
- #9389 (#208) Indians mounted and foot
- #9390 (#272) Cowboys and Indians mounted and foot

Second Grade Paint Series Cowboys and Indians came boxed as various numbered sets with an Alpha Suffix including A, C, N, P, S, and W.

North American Indian Chief with knife. Picture Pack #22B. A second grade paint version was allocated Master Model #597B (not illustrated). $15-$20 each. Picture Pack, $25-$30.

North American Indian Chief/Medicine Man with tomahawk. Picture Pack #23B. The version carrying a knife was included only in some pre-War boxed sets. $15-$20 each. Picture Pack, $25-$30.

North American Indian Chief. Fixed arm with tomahawk, feathered headdress. Picture Pack #24B. See also second grade version #99P. $10-$15 each. Picture Pack, $20-$25.

Cowboy, large brimmed hat walking with rifle. Picture Pack #25B. $15-$20 each. Picture Pack, $25-$30.

North American Indian Chief, galloping horse on stand, carrying rifle. Picture Pack #33B. $25-$30 each. Picture Pack, $35-$40.

North American Indian Chief mounted (no stand) with tomahawk. Picture Pack #32B. $25-$30 each. Picture Pack, $35-$40.

Cowboy mounted (no stand), with pistol. Picture Pack #34B. $25-$30. Picture Pack, $35-$40.

North American Indian Chief mounted (no stand) with rifle. Master Model #49B. This is the same as #32B with an arm change and was not issued in the Picture Pack range. $25-$30 each.

Cowboy mounted with lasso, rearing horse with stand. Picture Pack #35B. $30-$35 each. Picture Pack, $40-$45.

Cowboy mounted with pistol on rearing horse. Master Model #38B. Not issued in the Picture Pack Range. $30-$35 each.

Cowboy on foot walking with pistol, moveable arm. Picture Pack #55B. $15-$20 each. Picture Pack, $25-$30.

North American Indian Brave with rifle at the ready (old model). Picture Pack #74B. Issued in boxed sets but without a Master Model #. This item is more commonly associated with Catalog #17N or as second grade item #51P. $10-$15 each. Picture Pack, $20-$25.

Cowboy mounted with lasso (no stand). Master Model #106B. $30-$35 each. Picture Pack, $40-$45.

North American Indian Brave with rifle at the ready. Picture Pack #74B also. This is a newer design and replaced the old #17N. $15-$20 each. Picture Pack, $25-$30.

North American Indian Chief galloping horse on stand with hatchet. Master Model #169B. This is the same as #33B with an arm change and was not issued in the Picture Pack range. $25-$30 each.

Cowboy crouching firing pistol, fixed arm. Picture Pack #274B. See also second grade version #80P. $10-$15 each. Picture Pack, $20-$25.

Cowboy walking with lasso. Picture Pack #356B. Also issued in the pre-War Circus range with a sand colored base. $25-$30 each. Picture Pack, $30-$35.

North American Indian crawling with knife, moveable arm. Picture Pack #1179B. The second grade fixed arm version was allocated Catalog #55P. $15-$20 each. Picture Pack, $25-$30.

Cowboy on foot walking with pistol, fixed arm. Not issued as a Picture Pack. Master Model #599B. See also second grade paint version #83P. $10-$15 each.

Cowboy mounted with rifle, no stand. Picture Pack #1180B. $25-$30. Picture Pack, $35-$40.

Frontiersman seated with rifle, Master Model #1210B. This item was the seated aged villager figure from the Farm Series, Catalog #555. Cowboy seated with plain arm using Farm, Catalog #560 tractor or horse drawn vehicle driver, Master Model #1213B. Cowboy sitting with pistol in moveable arm, Master Model #1220B. Cowboy sitting with Stetson in moveable arm, Master Model #1221B. Left to right: $50-$55, $30-$35, $30-$35, $80-$90, $80-$90.

Cowboy on bucking horse, bronco. Picture Pack #1219B. $80-$90. Picture Pack, $100-$120.

Artist's impression of Cowboy on bucking horse, not adopted by Britains.
Courtesy of Arnold Rolak.

North American Indian crouching with knife and hatchet. Second grade paint version of #143B issued as a Picture Pack (not illustrated). Catalog numbers 52P, 16N, 16P and 35C. $8-$10 each.

North American Indian Brave with rifle at ready. Second grade paint version of #74B. Catalog numbers 51P, 17N, 17P, 25C, and 4A. $8-$10 each.

North American Indian crawling with knife, fixed arm. Second grade paint version of #1179B. Catalog numbers 55P, 15N, 15P, 27C and 4Q. $8-$10 each.

Cowboy crouching with pistol. Second grade paint version of 274B. Catalog numbers 80P, 4N, and 4P. $8-$10 each.

North American Indian Chief with tomahawk, fixed arm. Second grade paint version of #24B. Catalog numbers 99P, 14N, 14P, and 54C. $8-$10 each.

Top: Cowboy walking with pistol, fixed arm. Second grade paint version of #599B. Catalog numbers 83P, 5N, 5P, and 30C. $8-$10 each.

Bottom: Cowboy standing firing pistol. Second grade paint version of #275B issued as a Picture Pack (not illustrated). Catalog numbers 96P, 6N, and 6P. $8-$10 each.

North American Indian Chief mounted with tomahawk, fixed arm. Second grade paint. Two versions exist with horses tail up or down. Tail up version #66N and #78C. Tail down version #117P, #63N and #35C. $10-$15 each.

Cowboy mounted firing pistol backwards. Second grade paint. Catalog numbers 118P, 64N, and 36C. $10-$15 each.

Indian swinging club, catalog #155P in the Crown range; issued also in 1953 it was allocated Master Mold #1418. No record exists of this item's Master Model #. *Courtesy Stewart Saxe.* $60-$75 each.

North American Indian Chief carrying rifle, #136P. Second grade paint version of Picture Pack #1216B (not illustrated) and North American Indian Chief crouching with rifle, #137P. Second grade paint version of Picture Pack #1217B (not illustrated). $10-$12 each. Picture Packs, $30-$35 each.

Very rare Cowboy standing firing pistol introduced into the Crown range as Catalog #154P. Second grade paint figure in 1953 allocated Master Model #1472B and Master Mold #1417. This piece also appeared in #2145 half-set in first grade paint. *Courtesy Stewart Saxe.* $60-$75 each.

Cowboys walking with pistol showing "D" Series size and 54mm standard size. $10-$15 each.

Cowboys and Indians "D" Series small mounted Cowboy #22D, Cowboy on foot #23D, Indian Brave #21D, and mounted Indian #20D. Foot, $10-$15 each. Mounted, $15-$25 each.

American Indians. Post-War box set #150. *Photo courtesy Bob Bailey.* $120-$130.

American Heroes set. Britains supplied the Cowboy and Indian figures to Beiser in the U.S.A. for this shooting game, which came supplied with a cork pop gun. *Photo courtesy Ray Haradin.* $1,400-$1,600.

Cowboys mounted. Pre-War box set #179. *Photo courtesy Bob Bailey.* $130-$150.

Cowboys mounted. Post-War box set #179. $110-$130.

American Cowboy. Pre-War box artwork, set #183. *Photo courtesy Bob Bailey.* $120-$130.

Cowboys on foot. Post-War box set #183. $120-$140.

Cowboys and Indians display set. Post-War box set #208 containing Indians. *Courtesy Bob Bailey*. $130-$150.

American Cowboys. Pre-War box set #245. *Photo courtesy Bob Bailey*. $130-$150.

Cowboys and Indians display. Post-War box set #272. *Photo courtesy Bob Bailey*. $110-$130.

Cowboys and Indians display set. Post-War box set #209 containing Cowboys. *Photo courtesy Bob Bailey*. $130-$150.

Cowboy and Indian display. Pre-War box set #273. *Photo Courtesy Ray Haradin*. $1,200-$1,400.

Wild West box art for set #273. *Courtesy Ray Haradin*.

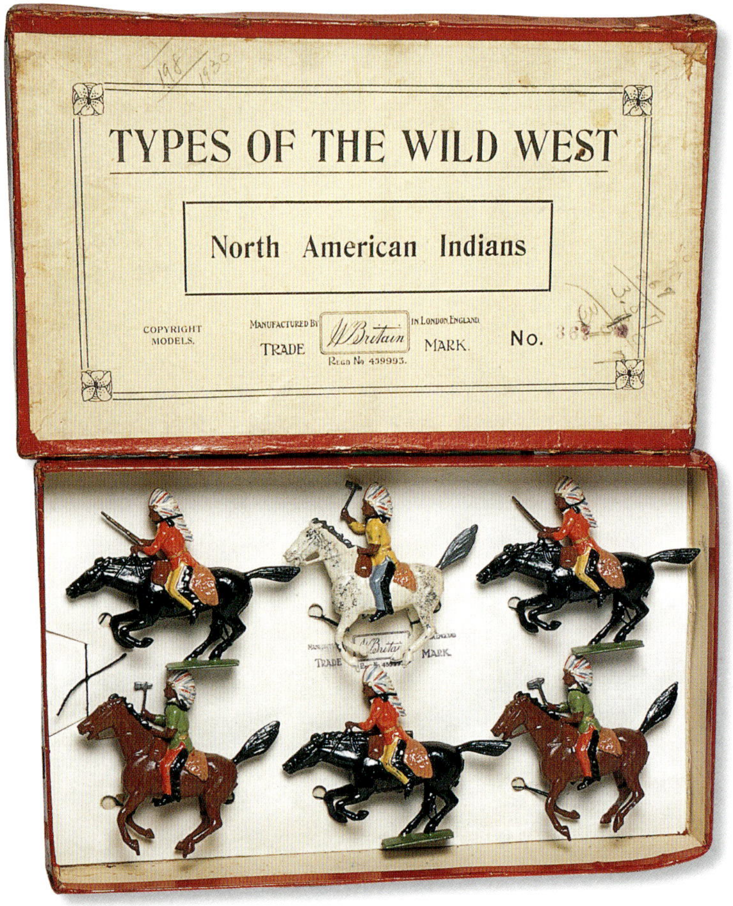

Types of the Wild West. Pre-War box set #367 containing mounted Indians. A previously undiscovered non-cataloged set. *Photo courtesy Arnold Rolak*. $300-$400.

Texas Rangers. Pre-War set #1508. This set contained the standard mounted cowboys with a label showing "stills" from the movie of the same name starring Fred MacMurray and released in 1937. *Courtesy Christie's Auction House, London/The Arnold Rolak Reference Collection*. $4,500-$5,000.

Texas Rangers. Box art for larger set #1509.

Prairie Schooner with canvas top issued in both sets #2034 and #2042. The Schooner was allocated Master Model #1193B. $150-$180.

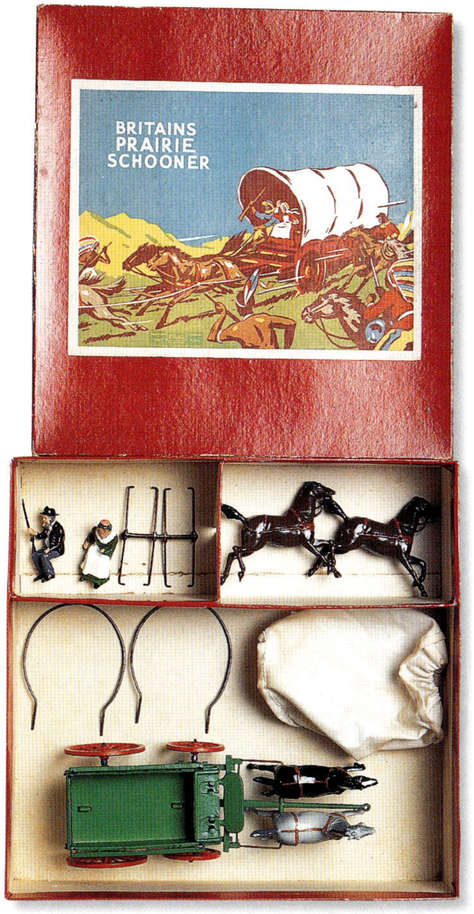

Prairie Schooner. Post-War box set #2034. *Courtesy Philip Dean Reference Collection*. $750-$800.

Prairie Schooner with white tin top issued in both sets #2034 and #2042. $150-$180.

Prairie Schooner with rare brown tin top. *Photo courtesy of The Blue Ox.* $180-$200.

Set #2042. Prairie Schooner with Cowboys and Indians box art. *Courtesy Philip Dean Reference Collection.*

Set #2042. Prairie Schooner with Cowboy escort and attacking Indians. *Courtesy Philip Dean Reference Collection.* $900-$1,200.

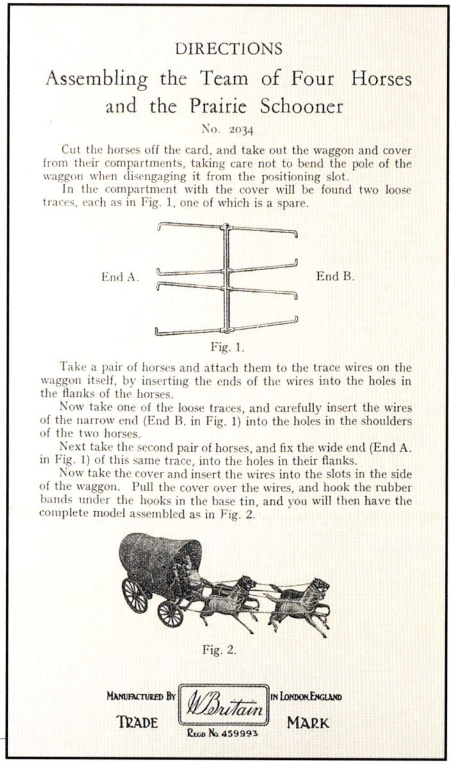

DIRECTIONS

Assembling the Team of Four Horses and the Prairie Schooner

No. 2034

Cut the horses off the card, and take out the waggon and cover from their compartments, taking care not to bend the pole of the waggon when disengaging it from the positioning slot.

In the compartment with the cover will be found two loose traces, each as in Fig. 1, one of which is a spare.

End A. End B.

Fig. 1.

Take a pair of horses and attach them to the trace wires on the waggon itself, by inserting the ends of the wires into the holes in the flanks of the horses.

Now take one of the loose traces, and carefully insert the wires of the narrow end (End B. in Fig. 1) into the holes in the shoulders of the two horses.

Next take the second pair of horses, and fix the wide end (End A. in Fig. 1) of this same trace, into the holes in their flanks.

Now take the cover and insert the wires into the slots in the side of the waggon. Pull the cover over the wires, and hook the rubber bands under the hooks in the base tin, and you will then have the complete model assembled as in Fig. 2.

Fig. 2.

Manufactured By W. Britain In London England
Trade Mark
Regd. No. 459993

Interior of box lid for set #2042 showing assembly instructions for the Prairie Schooner traces. *Courtesy Philip Dean Reference Collection.*

Rodeo set. Post-War box set #2043. *Courtesy Philip Dean Reference Collection.* $1,200-$1,500.

Rodeo set. Example of a displayed set. Master Model numbers were given to the components of the corral: Fencing side feet, #1222B; Fencing corner fee, #1223B; long wooden dowels (poles), #1224B; short wooden dowels (poles), #1225B.

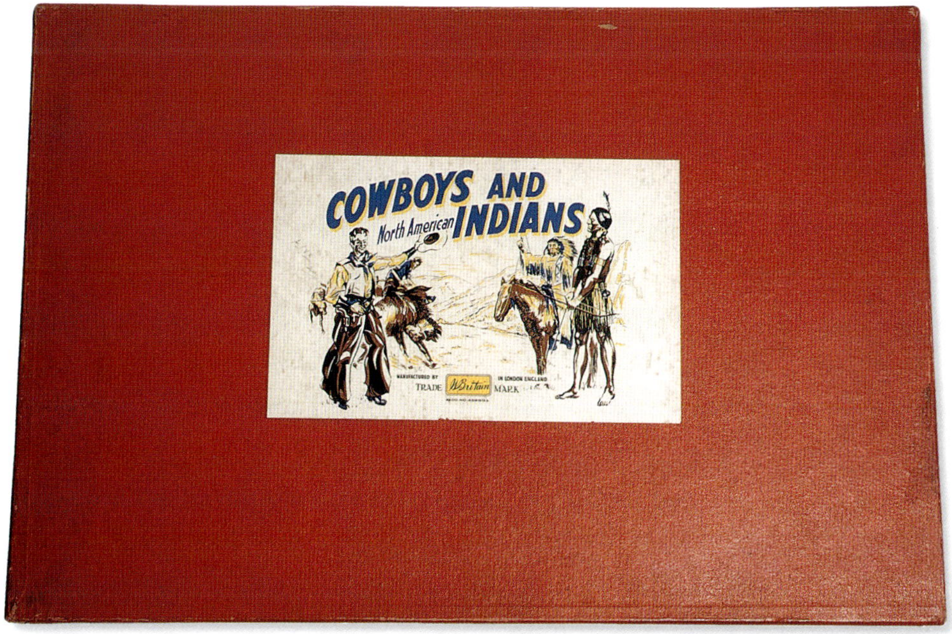

Cowboy and Indian display set #2061, box art. This was the second largest post-War set made by Britains, Ltd. It contained a lift-out upper tray. Plastic fencing was included and it was one of the first transitional sets issued after the Herald Company was acquired by Britains, Ltd. *Photo courtesy of Bob Bailey*. $3,500-$4,000.

Set #2061, upper tray. *Photo courtesy Bob Bailey.*

Set #2061, lower tray. *Photo courtesy Bob Bailey.*

Box art for half-sets #2145 and #2146. *Courtesy Stewart Saxe.*

Mazak Cowboy and Indian set #2163Q. Post-War issued for the Australian market. *Photo courtesy Bob Bailey*. $600-$800.

American Indians, half-set #2146. A similar set of Cowboys was also issued as #2145 (not illustrated). $150-$180.

Mounted Indians. Post-War cellophane-fronted box set #9289. Formerly set #152. *Photo courtesy Bob Bailey.* $110-$120.

Cowboy display set. Post-War cellophane-fronted box set #9388. Formerly set #209. *Photo courtesy Bob Bailey.* $110-$120.

Indian display set. Post-War cellophane-fronted box set #9389. Formerly set #208. *Photo courtesy Bob Bailey.* $110-$120.

Picture Pack box art work showing Wild West scenes.

Cardboard/tin which contained second grade Cowboys and Indians. Artist's impression of the contents appear on the cardboard label. $60-$70.

Cowboys and Indians. Second grade display set. Catalog #44S. *Photo courtesy Bob Bailey*. $150-$180.

Cowboys and Indians. Second grade box set with caricature lid. Catalog #164S. *Photo courtesy Bob Bailey.* $40-$60.

Cowboys and Indians. One of 30 sample sets in 1988. Made of die cast lead from original second grade hollow cast molds. Britains decided not to issue this item. *Photo courtesy Arnold Rolak/Christie's Auction House, London*. $300-$350.

Second grade "S" Series. Wild West display boxes, post-War issue. *Courtesy Stewart Saxe*. $60-$80 each.

Second grade "S" Series. Large display set containing sixteen Cowboys and Indians, post-War issue. *Courtesy Stewart Saxe*. $100-$120.

Second grade "S" Series. Cowboy and Indian set, post-War issue. *Courtesy Stewart Saxe*. $60-$80.

Box art for large "S" Series. Cowboy and Indian set. Post-War issue. *Courtesy Stewart Saxe.*

Box art for double row. Cowboy and Indian "S" Series. Second grade figures, post-War issue. *Courtesy Stewart Saxe.*

Box art. Rainbow Toys, a trade name used by Britains for second grade "S" Series figures, post-War issue. *Courtesy Stewart Saxe.*

Box art. Early post-War label box for "S" Series single row. Cowboy and Indian Series. *Courtesy Stewart Saxe.*

10. Boys Scouts: 1910

The Scout movement formed by Baden Powell obviously influenced the company's decision to issue Boy Scout sets in 1910. The Signalers and a Scout Encampment were the only sets to survive post-War. Girl Guides and Guider and an American Girl Scout were introduced into the catalog during the 1930s. Combination sets containing Boy Scouts and U.S. Girl Scouts were issued under set numbers 289 and 293. A cloth Boy Scout tent came with set #1761. However, these items, because of their extreme rarity, have not become available for photography.

Scout Master, Master Model #15B. Pre- and post-War paint variants. $65-$85 each.

Boy Scout kneeling with axe, Master Model #16B; Boy Scout with two signal flags in moveable arms, #797B; Boy Scout walking with pole, #799B; Boy Scout walking with pole on shoulder, #798B; Boy Scout with pennant on pole, #818B. $55-$65 each.

Boy Scout with trek cart. Pre- and post-War color variants. Scout is Master Model #825B, Trek cart, #827B. $150-$180 set.

Trek carts, Pre- and post-War versions, Master Model #827B. The rails detach and clip together to form ladders (see Display Sets). The long rail was Master Model #826B and the short rail, #828B. $80-$100 each.

Boy Scout "Lookout," Master Model #819B; Boy Scout plain arms, #821B; Boy Scout standing with axe and hooked hand to hold on to tree branch or fence, #823B. $55-$65 each.

Boy Scouts with horizontal poles, fixed arm, second grade paint figures. Catalog #IN. $40-$45 each.

Boy Scout set #161 box art. This set contained a Scout Master and eight Boy Scouts with poles. Pre-War issue only.

Set #163, Boy Scout signallers. Post-War example. The semaphore flags, when positioned as in the box lid diagram, spell out the word Scout. Issued also pre-War. $400-$600.

Set #180 Boy Scout's display issued Pre- and post-War. Some of the Scouts in the set illustrated have a very unusual basic paint style (See top row). *Courtesy Ray Haradin.* $2,000-$2,200.

Set #181 box art, Boy Scouts display. *Courtesy Ray Haradin.*

Set #181 Boy Scout's display combining the contents of set #162 (not illustrated) and set #180. This 45 piece set was issued both Pre- and post-War. *Courtesy Ray Haradin.* $6,000-$7,000.

Very rare Boy Scout signallers in blue uniforms (Sea Scouts?). Purchased in Paris and possibly made at the French factory. *Courtesy Dr. Rob Wilson.*

Girl Guides and Girl Scouts

Set #238 American Girl Scouts. *Courtesy Ray Haradin.* $3,500-$4,000.

American Girl Scout Master Mold #17B. *Courtesy Philip Dean Reference Collection.* $200-$250.

Girl Guides with Guider from set #1332. See Farm Chapter for Master Model numbers. *Courtesy Philip Dean Reference Collection*. Price: See Farm Chapter.

Left: Girl Guide Mistress and Girl Guide. Both allocated #607 in the Farm numbering system; See Farm Chapter for Master Model numbers. *Courtesy Philip Dean Reference Collection*. Price: See Farm Chapter.

Right: American Girl Scout, Master Model #17B and Girl Guide Mistress, Catalog #607 shown for comparison. *Courtesy Philip Dean Reference Collection*. $90-$110.

11. Farm People: 1922

Farm figures, animals, accessories and the first horse-drawn farm items were first produced in 1921, but were advertised the first time in the 1922 catalog. Prior to this, and possibly as early as 1911, some trees and shrubs were included in various military, Boy Scout, and Cowboy and Indian sets. In addition, an unsaddled young standing horse, dated August 15, 1903, has also been found (not illustrated; see *The Great Book of Hollowcast Figures,* by Norman Joplin, published by New Cavendish Books).

The term Model Home Farm is attributed mainly to items issued pre-World War Two. On resumption of production after World War Two, the range became known simply as Britains Farm.

The consumer's apparent rejection of military style and war toys after World War Two seems to have been a major contributing factor in Britains decision to issue this farm and other non-military ranges which have collectively become known as "civilians."

Farm Figures Itemized with Date of First Issue

Catalog#	Issue Date	Catalog#	Issue Date	Catalog#	Issue Date
501-529	1922	756	1949	90F-92F	1938
530-533	1923	758	1950	94F-111F	1939
536-541	1925	769	1951	112F-113F	1940
542-566	1926	775-779	1954	120F-125F	1947
567-587	1927	780-781	1956	126F-129F	1948
588-607	1928	782-786	1955	130F-133F	1949
608-623	1929	1F-4F	1922	134F-137F	1950
624-634	1930	5F	1924	138F-139F	1951
635-638	1931	6F-11F	1926	142F	1952
639-641	1932	12F-17F	1928	144F	1954
642-647	1933	18F-20F	1929	145F-157F	1955
648-652	1934	21F-22F	1930	171F-174F	1960
653	1935	27F-30F	1932	175F-176F	1961
659	1936	31F	?	500F-505F	1961
660	1937	36F-45F	1933		
663-664	1939	51F-61F	1934	**Picture Packs**	
666-680	1940	62F-67F	1935	5001-5033	1948
715	1947	69F-74F	1936	5034-5036	1950
744-747	1948	77F-80F	1937	5037-5039	1952

Tinplate Farm Advertising Sign. $100-$150.

#501 Farmer. Thick and thin arm casting versions were available. $20-$25 each.

#502 Farmer's Wife with basket. $20-$25 each.

#503 Farmer's Wife with basket and umbrella. The center figure shows a pre-War "Made in England" tag necessary for importation from the United Kingdom to the USA in the 1930s. $20-$25 each. With tag $30-$35.

#505 Carter with whip. $20-$25 each.

#504 Carter. Plain arm version of #505. $20-$25 each.

#531 Milkmaid carrying pail on head. Pre-War issue only. $25-$30 each.

#532 Milkmaid carrying bucket by side. $20-$25 each.

#535 Landgirl. Pre-War issue only. The figure shown on the right has been given a post-War paint finish and was possibly one of a number of pre-War castings painted and issued post-War. This item is not included in any Britains post-War catalogs. $20-$25 each. Post-War paint variant, $40-$45.

#537 Milkmaid sitting on stool with milking pail. $20-$25 each.

#547 Man pushing wheelbarrow. See #564. $25-$30 each.

#554 Farmer's son sitting. Pre-War issue only. $35-$40 each.a

#555 Aged villager man sitting with pipe. This item was adapted and given a rifle arm when used in the Wild West Prairie Schooner set. $20-$25 each.

#556 Aged villager woman sitting. The figure on the right has a pre-War export tag (see #503). This item was also used as a pioneer woman in the Prairie Schooner set. $20-$25 each.

#557 Village girls. Pre-War issue only. The three items on the right have a post-War paint finish. See #535. $60-$70 each. Post-War variants $80-$90 each.

#558 Village boy with stick. $55-$65 each.

#559 Young lady walking. Also issued in the Railway and Civilian Series sets #158 and #168. The figure on the right was painted to represent a tennis player and was issued to the Australian market to compliment a wooden tennis court available from Australian department store, Walther and Stevenson. The item was allocated individual #806 in the Railway Series. $25-$30 each. Tennis Player variant, $80-$100.

#560 Farm hand seated for driving horse-drawn and motorized farm vehicles. Also issued as a Cowboy in the Rodeo set. $20-$25 each.

#561 Farmer's daughter sitting. Pre-War issue only. The two figures on the right have a pre-War export tag (see #503). $35-$40 each.

#562 Golfer. Pre-War issue only. $80-$90 each.

#563 Stable lad. $20-$25 each.

#564 Man with swing water barrow and barrel (see #547). $30-$35 each.

#577 Shepherd with crook. $25-$30 each.

#578 Automobile Scouts - A.A., walking and saluting. Both issued pre-War only to this number. Master Mold #179 is allocated taso the walking Scout and #180 is allocated to the saluting Scout. $80-$90 each.

#578 Automobile Scouts - A.A.. Color variants. See also Road Chapter for boxed set and Farm Accessories for A.A. signs. A rare version of the walking figure painted in a blue uniform to represent R.A.C. also exists. See *The Great Book of Hollow Cast Figures* by Norman Joplin, published by New Cavendish Books. $85-$90 each. Yellow sash variant, $120-$150.

#587 Village Idiots. Pre-War issue only. *Courtesy Philip Dean* Left to right: $300-$350, $300-$350, $240-$260, $300-$350, $240-$260, $240-$260, $300-$350, $300-$350, $240-$260, $240-$260.

#587 Village Idiot. Close-up of face. *Courtesy Philip Dean.*

#587 Village Idiots. Pre-War issue only. *Courtesy Philip Dean.* Left to right: $300-$350, $240-$260, $240-$260.

#587 Village Idiots. The blue and green smocks have a post-War paint finish although this item was not issued after World War Two. Paint variants influence the price with the rarer colors demanding a premium. *Courtesy Philip Dean*. Left to right: $300-$350, $300-$350, $300-$350, $240-$260, $300-$350.

#592 Curate and #593 Country Clergyman. #592 often nicknamed the "Thin Vicar" was issued pre-War only, while #593 was illustrated as an artist's impression in pre-War catalogs with an umbrella. No such example has so far come to light. Left: $60-$75; right: $30-$35.

#589 Blacksmith with anvil. See #650 and #651. $50-$55 pair.

#591 Dairyman with yoke and pails. Man only allocated Master Mold #191. $45-$55 each.

#594 Shepherd carrying lamb. $25-$35 each.

#595 Shepherd boy with lantern. $45-$55 each.

#598 Gentleman farmer mounted. Also issued as part of Hunting Series. $65-$75 each.

#600 Boy on Shetland pony. The figure on the right has a pre-War export tag attached. $55-$65 each. Tag variant, $65-$70.

#618 Boy and Girl on seesaw. $80-$120.

#619 Very rare boxed example of boy on swing. $300-$400.

Paint variants of Boy for swing and Boy and Girl for seesaw. *Photo courtesy Christie's Auction House, London/Britains Archive 1994*. Master Mold #219 allocated to Boy and #220 allocated to Girl. The log was assigned Master Mold #218. $20-$30 each.

#619 Boy on swing. Master Mold #222 allocated to boy. $75-$85.

#621 Traffic Policeman. Also issued in boxed Road Sign set. $25-$35.

Opposite page;
Top left: #600 Boy on Shetland pony. Color variants. $55-$65 each.

Top right: #607 Girl Guide and Guider. Both figures issued pre-War only with this number. Master Model #955B allocated to Guide and #956B to Guider. Guide, $150-$180. Guider, $75-$85.

Bottom: #607 Guider and Girl Guides. Issued as set #1332. Pre-War only issue. *Courtesy Alf Mussett*. $1,200-$1,500.

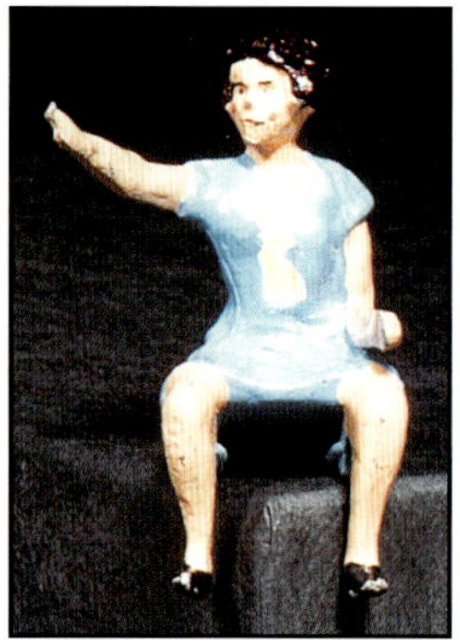

Motorcycle Pillion Passenger. Master Mold #246 allocated to girl and #245 allocated to driver. See previous photo. *Photo courtesy of Christie's Auction House, London/Britains Archive 1994*. $150-$200.

#641 Motorcycle and sidecar. Pre-War issue only. Master Mold #244 allocated to cycle, #247 to sidecar, and #248 to sidecar wheel. $800-$1,200 each.

#646 Navvy standing with shovel. $45-$55.

#645 Navvy with pick axe. $45-$55.

#650 Blacksmith. See #589 and #651. $25-$30.

#653 Man on motorcycle. Pre-War only issue. For Master Mold numbers, see Motorcycle and Sidecar. $600-$650.

#659 Policeman in peaked cap. $55-$65.

#652 Milk rounds man. Britains Factory records indicated that the arm holding book used for the Country Clergyman #593 was also used for this item. However, no example has so far come to light. $45-$50.

#678 Man to push wheelbarrow (see #547 and #564). #679 Man for lawnmower. Pre-War issue only and issued also in the Garden Series as #051 with #050 lawnmower. #680 Man for roller and #715 Man with garden roller. Post-War combination only. Left to right: $20-$25, $80-$100, $20-$25, and $45-$55.

#744 Farmhand sowing seed. Post-War only issue. $40-$45 each.

#745 Woman's Land Army. Post-War issue only. $40-$45.

#747 Girl with feeding bucket. Post-War only issue. $45-$55 each.

Below: #775 Mounted Policeman issued post-War only in Farm Series. Pre-War, this item was available in Police sets #319, #1511, and Coronation display set #1477. It was issued also on a brown horse in these sets. $75-$85.

#747 Girl with feeding bucket showing example with pre-War export tag. As this is a post-War only issue, it is not known why this figure has a pre-War tag. One theory is that the item may have been intended to have been issued pre-War, but was withheld. The figure is painted in post-War style. $100-$150.

#776 Policeman in helmet. Issued post-War only in Farm Series. Available pre-War as #808 in Railway Series and in Coronation display sets with tan base. #776 Policeman in helmet showing different color arm bands, red and white indicating City of London Police. $30-$35. Red and white armband variant, $70-$80.

#780 Painter to stand on ladder with paintbrush. Post-War issue only. #781 Painter to carry ladder #779. Post-War issue only. $75-$85 each. Ladder, $45-$55.

This Farmer's Wife has a fixed arm as opposed to the moveable arm #502 and #503 examples and could be a prototype. $60-$80.

Bottom left: Man to push plough. Not issued as a single item. Master Mold #100 allocated to this single piece. $35-$45.

Bottom center: #787 Garagehand. Also listed in catalogs as Handyman. Post-War issue only. $75-$80 each.

12. Farm Animals, 1922

#506 Shire Horse. $15-$20 each.

#507 Shire Colt. The figure on the right has a pre-War export tag attached to comply with U.S. Customs regulation in force during the 1930s. $10-$12 each. Tag variant, $12-$15.

#508 Cow standing. $12-$15 each.

#510 Sheep walking; #511 Sheep feeding; #512 Lamb. $8-$10 each.

#509 Calf. The item on the left is the first version casting. Master Mold #107 allocated to the first version and #113 to the second version. First version, $10-$12. Others, $8-$10.

#511 Sheep walking and feeding, first versions. *Courtesy Rex George*. $8-$10 each.

#513 Collie Dog. First version with one ear down. Master Mold #111 allocated to this first version. $15-$20 each.

#513 Collie Dog. Second version, both ears up. Master Mold #900 allocated to second version. $10-$12 each.

#514 Pig feeding. $8-$10 each.

#515 Turkey. $8-$10 each.

#516 Speckled Fowls; #517 White Fowls; #518 Yellow Fowls. All three use the same casting with different paint finishes. #517 and #518 are pre-War only; later included post-War as the same catalog #516. $8-$10 each.

#519 Angry Gander and #520 Goose. $8-$10 each.

#530 Sheep lying. $8-$10.

#533 Ducks and Drakes. Both issued with this number. $8-$10 each.

#534 Calf lying. A black and white color variation was also issued. See Picture Pack #5017 photo. $8-$10 each.

#536 Sheep and Lamb lying down. $10-$12 each.

#538 Cow lying. $12-$15 each.

#539 Cow feeding. $12-$15 each.

#540 Goat. Also used in Eastern People sets. $10-$12 each.

#541 Cart Horse. Two different castings were allocated this number both with collar and saddle, right foreleg forward. $15-$20 each.

#541 Cart Horse. Variant left foreleg/hoof raised. $15-$20 each.

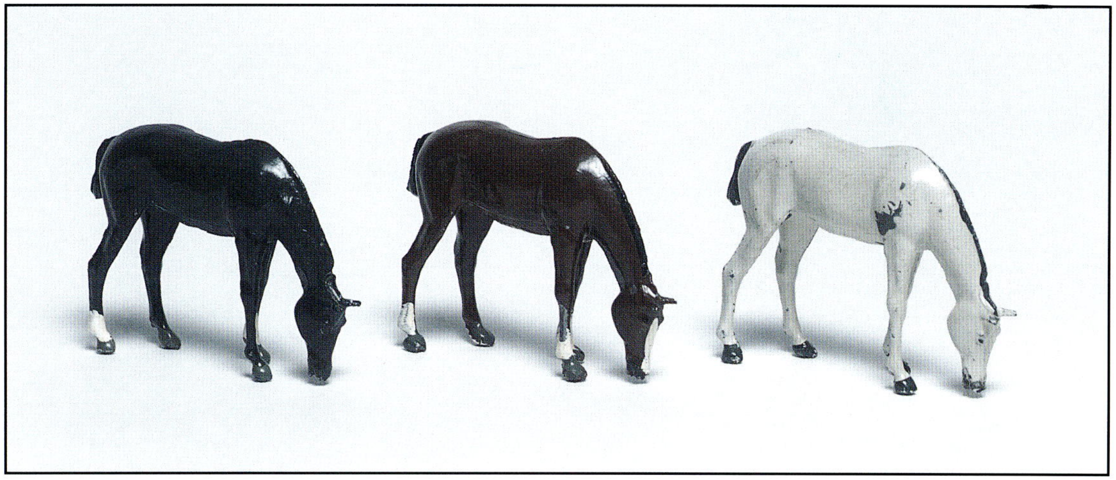

#543 Horse Feeding. $10-$15 each.

#544 Chicks. Running, pecking and flapping wings. All three allocated this number. $12-$15 each.

#545 Hen sitting. $8-$10 each.

#546 Piglets, head up and head down, feeding. Both allocated this number. $8-$10 each.

#550 Cob. $10-$15 each.

#552 Donkey standing. Also available in the Eastern People set. Master Mold #295 allocated to standing version. $12-$15 each.

#552 Donkey walking. Did not appear in any catalog. Master Mold #152 allocated to walking version. $30-$40.

#565 Goslings. Two different castings allocated to this number. Pre-War only issue. $10-$12 each.

#571 Dog lying. #572 Dog sitting. $10-$12 each.

#573 Bull. Pre-War tail up version. $25-$30.

#573 Bull. Post-War tail down and cast to body. $20-$25.

#576 St. Bernard dog. $10-$15.

#596 Berkshire Pigs, Boar and Sow. Both allocated to this number. Master Mold #196 allocated to Boar and #197 to Sow. $12-$15 each.

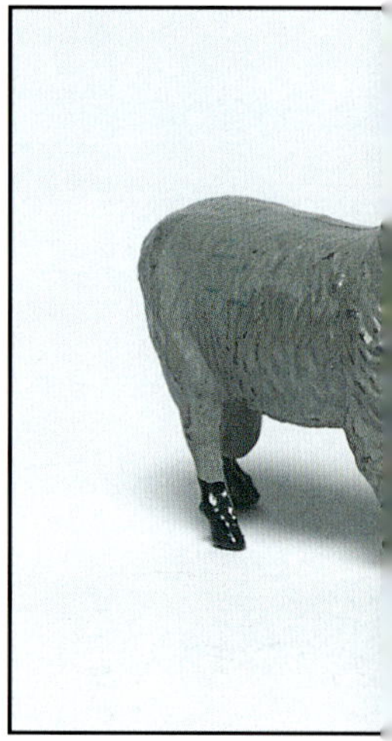

#597 Exmoor Horn Sheep, Ram and Ewe. Both allocated to this number. Master Mold #199 allocated to Ram and #198 to Ewe. $12-$15 each.

#599 Jersey Cow. The Jersey Cow pictured right has a bell and did not appear in any catalog. Issued post-War. Master Mold #200 allocated to Cow without bell and #1034 to Cow with bell. $20-$25. Variant with bell, $80-$100.

#602 Foal. $10-$12 each.

#601 Hampshire Down Ram. Sometimes nicknamed "dipped' sheep because of its orange color. The ram on the right has a pre-War export tag. $12-$15 each. Tag variant, $15-$20.

#603 Rabbit sitting. $12-$15 each.

#604 Cat seated. $15-$20 each.

#605 Greyhound standing. The figure on the right has a pre-War export tag. $25-$30 each. Tag variant, $30-$35.

#606 Greyhound running. The figure on the right has a pre-War export tag. $25-$30 each. Tag variant, $30-$35.

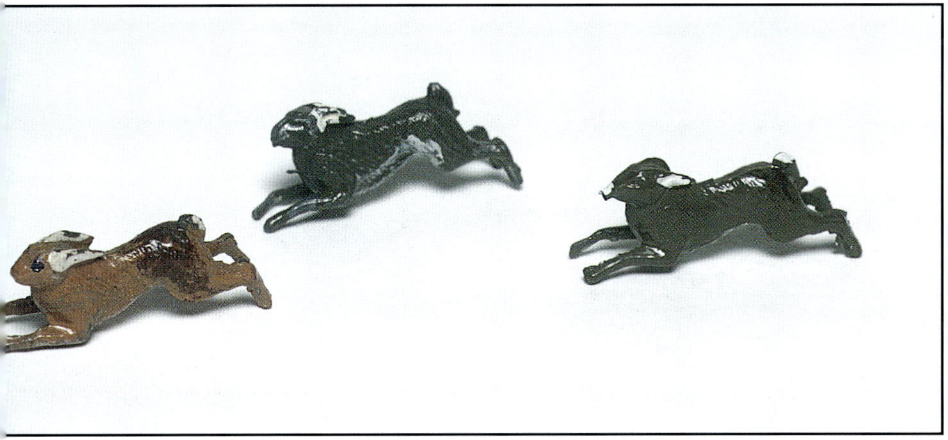

#620 Hare running. $12-$15 each.

#622 Swan and five Cygnets. Both items allocated to this number. Master Mold #226 allocated to Swan and #227 to Cygnets. Archive paint sample cards list the Swan as Catalog #661 and the Cygnets as Catalog #662. $65-$75 set. Single swan, $12-$15. Single Cygnet, $10-$12.

#636 Rabbit sitting up on back legs. Pre-War only issue. $45-$55.

#637 Dog begging and #638 Spiteful Cat. Pre-War only issued. Dog, $75-$80; Cat, $85-$90.

#642 Rhode Island Red Prize Poultry. Cocks and Hens standing. Both allocated this number. Pre-War issue only. Master Mold #249 allocated to Cock standing and #250 to Hen standing. $20-$25 each.

#643 White Leghorn Prize Poultry. Cocks and Hens standing and feeding. All four casting allocated this number. Master Mold #249 allocated to Cock standing, #250 to Hen standing, #258 to Hen feeding, and #259 to Cock feeding. Master Mold #. $20-$25 each.

#644 Black Plymouth Rock Prize Poultry. Cock and Hen standing. Both allocated this number. Pre-War issue only. See previous illustration. #660 Prize Poultry feeding Hen and Cock, both issued to this number. Master Mold #259 allocated to Cock and #258 to Hen. $25-$35 each.

#647 Highland Cattle. $25-$35.

Highland Cattle Calf (right) not shown in any catalog. Appeared on paint sample card in Christie's Britain archive sale as number #645. Shown here for comparison with #534. Highland Calf, $40-$50. *Courtesy Brian Holmes.*

#648 Field Horse with collar. No saddle, left foreleg/hoof raised. $20-$25.

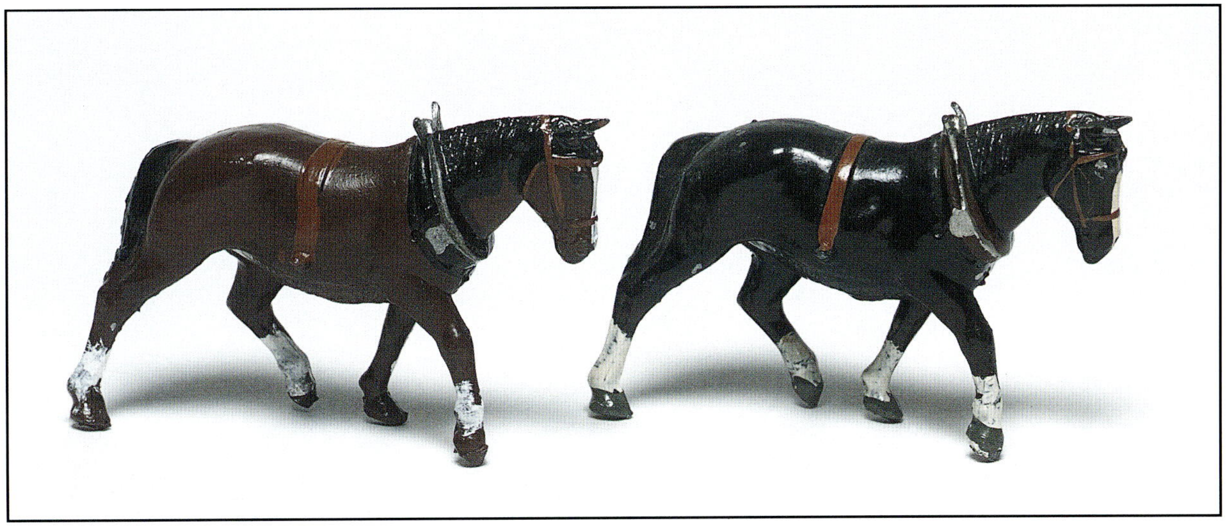

#649 Field Horse. Collar, no saddle, right foreleg forward. $20-$25.

#746 Berkshire Sow and Piglets. Post-War issue only. $65-$70.

#756 Wild Horse. Also issued with Rodeo and Wild West displays. Post-War issue only. $30-$35.

#756 Wild Horse, rare. White color variant. Post-War issue only. $80-$90.

#758 Bullock. A black and white paint variant was sold as part of the Britains archive sale at Christie's in 1994. This item was also issued in both Rodeo and Wild West displays. Post-War issue only. $30-$35 each.

#769 Field Horse. This is a re-cataloged #649 Field Horse and was introduced into the 1951 catalog at the same time as #648 and #649 were deleted. $20-$25.

#782 and #783 Suffolk Mare and Foal. Color variations. Post-War issue only. Mare, $85-$90. Foal, $75-$85.

#782 Suffolk Mare and #783 Suffolk Foal. Post-War issue only. Mare, $75-$85;, Foal, $70-$75.

#784 Ayrshire Bull, #785 Ayrshire Cow, and #786 Ayrshire Calf. Post-War only issues. Bull, $150-$180. Cow, $180-$200. Calf, $90-$120.

Cart Horse. Sturdy Clydesdale type found only in early version #12F Timber Trailer boxed sets. Nicknamed "Fur Hoofed" cart horse and referred to in the Master Mold list as #263. $60-$65.

#782 Suffolk Mare and #758 Bullock. Color variants. *Courtesy Christie's, London*. Bullock, $80-$100. Mare, $85-$90.

Example of Britains printed wrapping paper for bulk range items supplied to the trade.

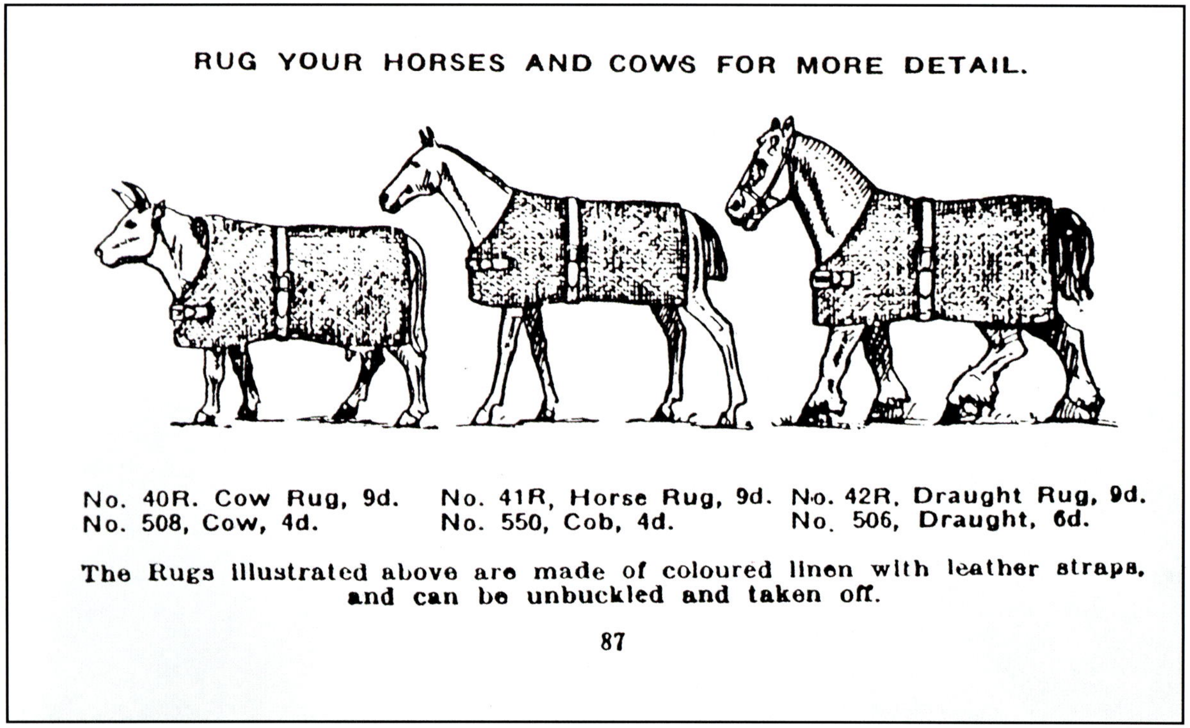

RUG YOUR HORSES AND COWS FOR MORE DETAIL.

No. 40R. Cow Rug, 9d. No. 508, Cow, 4d.

No. 41R, Horse Rug, 9d. No. 550, Cob, 4d.

No. 42R, Draught Rug, 9d. No. 506, Draught, 6d.

The Rugs illustrated above are made of coloured linen with leather straps, and can be unbuckled and taken off.

87

Walther and Stevenson, an Australian Department Store, manufactured and retailed rugs to fit Britains cows and horses made of colored linen with leather straps. These items are considered exceptionally rare. *Courtesy Robin Yates.*

13. Farm Accessories, 1922

#521 Oak Tree and, #522 Cedar Tree, pre-War only issues. #523 Elm Tree and #524 Fir Tree. *Courtesy Brian Holmes*. Left to right: $70-$75, $60-$65, $45-$55, $45-$55.

#525 Fallen Tree, pre-War only issue., #526 Shrub/Hedge and #527 Hurdle. Tree, $30-$35; Shrub, $15-$20; Hurdle, $10-$12.

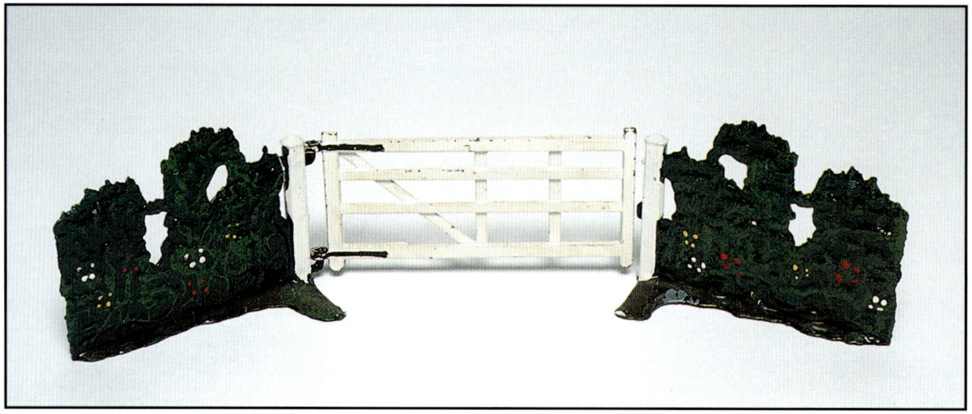

#548 Hedges and Field Gate. $55-$65.

#528 Large Trough, #529 Small Trough, and #542 Wheat Sheaves stacked. $15-$20 each.

#549 Hedges and Garden Gate, pre-War issue only. $75-$85.

#551 Scarecrow with removable hat. Master Mold #150 was allocated to the Scarecrow and #151 to the hat. $30-$35 each.

#553 Single Wheat Sheaf, pre-War issue only; #566 Field Hayrack; and #567 Log Seat. Single Wheat Sheaf, $30-$35; Hayrack, $15-$20; Log Seat, $10-$15.

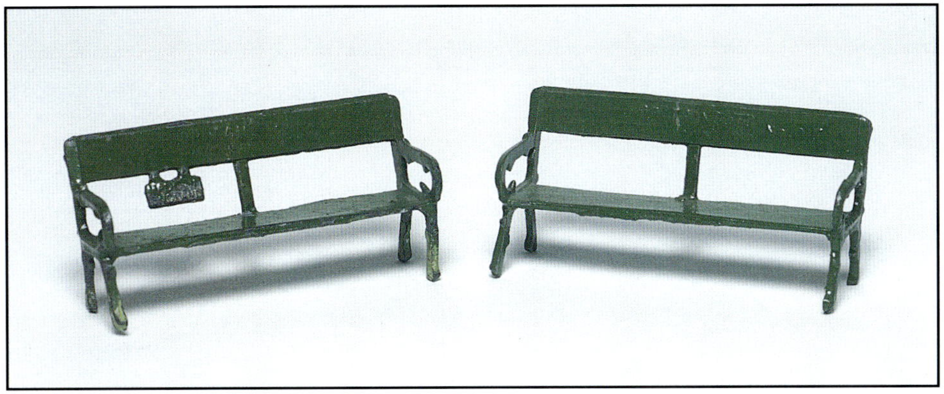

#568 Garden Seat. Left item has pre-War export tag. Also issued in Garden Series as #017 in both green and white. See Garden chapter for examples. $12-$15. Tag variant, $15-$20.

#569 Dog Kennel and #570 Dog Kennel with baseboard (pre-War issue only). Baseboard was allocated Master Mold #170. Dog, $10-$12, Kennel with baseboard, $70-$75.

#569 Dog Kennel and #570 Dog Kennel with baseboard with #571 Dog lying and #572 Dog sitting. Kennel, $30-$35; with Baseboard, $70-$75; Dogs, $10-$12 each.

#574 Telegraph Pole and #575 Dovecote. $35-$45 each.

#581 Rustic Stile; #582 Sign Post, one way; #583 Sign Post, two way; #584 Sign Post, three way. Although Catalog #585 was not allocated an item, it is believed this was left blank for the issue of a four way sign post of which no example has yet come to light. Stile, $15-$20, Sign Posts, $20-$40.

#579 A.A. Destination Sign, circular "Crawley." A.A. Caution signs, triangular "School" and "Dangerous Corner." Paper stickers were used on these sign posts. Master Mold #181 was used to facilitate the change of top from triangular to round. Pre-War issue only. $65-$80 each.

#617 Garden Table. Three Master Mold numbers were used for the table: #215 Table Top, #216 Outside Legs, #217 Inside Legs. $80-$90.

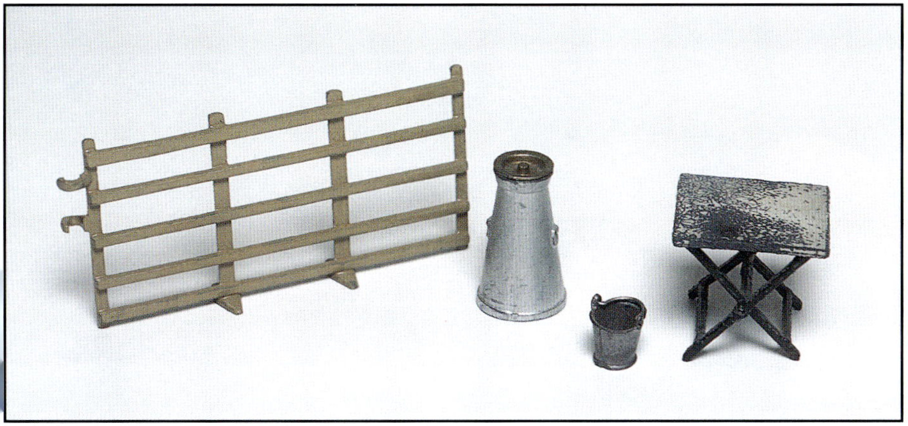

#586 Fencing, #588 Milk Churn. Master Mold #188 allocated to churn and #189 to lid. #590 Pail, #617 Garden Table. Fencing, $10-$12; Churn, $20-$25; Pail, $5-$10; Table, $80-90.

#624 Flint Wall straight section, #625 Flint Wall round corner section, #626 Stile for Flint Wall. Pre-War issue only. $60-$65 each.

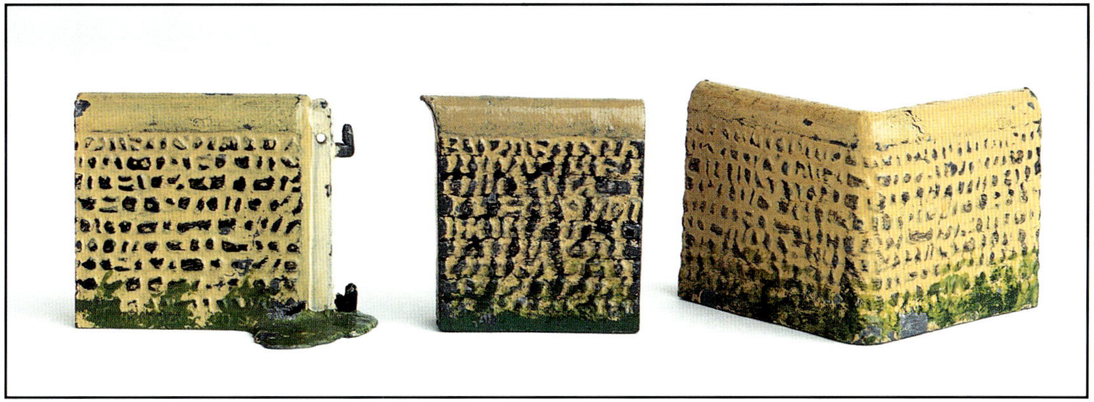

#627 Flint Wall gate post section, Flint Wall short cross section and #629 Flint Wall short corner section. Pre-War issues only. $50-$55 each.

#630 Five Barred Gate (large) and #631 Five Barred Gate (small). Pre-War issues only. $15-$20 each.

#632 Tryst Gate Frame, #633 Stone Pier, and #634 Stone Pier for Gate (#630 or #631). Pre-War issues only. Tryst Gate Frame, $80-$90. Others, $35-$45 each.

#639 Small Shrubs. Two different castings allocated to this number. Pre-War issue only. #640 Small Tree with round or "H" shaped base. $15-$20 each.

#635 Tinplate Pond with Swan and two Cygnets. Pre-War issue only. *Courtesy Philip Dean Reference Collection*. $250-$300.

#651 Anvil, #667 Garden Roller, two castings, one piece and two piece roller barrel. #670 Garden Wheelbarrow, #673 Lawnmower (pre-War issue only), and Swing Water Barrow (see #564 in Farm People section). Lawn Mower, $65-$75. Others, $10-$15 each.

14. Farm "F" Suffix, including Horse Drawn & Vehicles, 1922

Farm sets, Vehicles, Buildings, Accessories and Horse-Drawn Vehicles, and Equipment, all with the "F" suffix catalog number.

Farm Items, Pre-War Not Illustrated

#2F. Model Farm display set, 40 pieces.
#3F. Model Farm display set, 71 pieces.
#10F. Shepherd with Flock of Sheep boxed set.
#11F. Milkmaids and Cows boxed set.
#13F. Village Group boxed set.
#14F. Farmyard display boxed set.
#15F. Cattle display boxed set.
#16F. Stable display boxed set.
#17F. Farm display boxed set.
#18F. Farm display boxed set.
#27F. Farm Assortment boxed set.
#28F. Farm Assortment boxed set.
#29F. Farm Assortment boxed set.
#37F. Farm display boxed set.
#51F. Farm display boxed set.
#52F. Large Farm display boxed set.
#63F. Farm display boxed set.
#69F. Farm Assortment boxed set.
#71F. Farm Assortment boxed set.
#72F. Farm Assortment boxed set.
#73F. Farm Assortment boxed set.
#74F. Farm Assortment boxed set.
#77F. Farm Assortment boxed set.
#80F. Shepherd and Sheep set.
#90F. Builder's Lorry, 4 wheel. See #59F.
#91F. Builder's Lorry, 6 wheel. See #60F.
#92F. Builder's Lorry, 10 wheel. See #61F.
#95F. Large Barn and Cart Shop.
#96F. Stable, wooden.
#99F. Cow Shed, wooden.
#103F. Mansard Barn.
#111F. Farmyard presentation box.
#112F. Greenhouse, See Garden Section.
#113F. Garden Shelter, See Garden Section.

Post-War Only Issues, Not Illustrated

#121F. Farmyard display.
#132F. Farmyard display.
#139F. Tractor and Clockwork Trailer.
#145F. Tractor and Implements set.
#146F. Tractor and Implements set.
#148F. Farmyard display.
#149F. Farmyard display.
#151F. Farmyard display.
#153F. Farmyard display.
#154F. Farmyard display.
#157F. Farmyard display.
#171F. Metal Wheeled Fordson Power Major Tractor with power steering wheel adapted from #127F allocated #9526 in 9000 Series.
#172F. Rubber Wheeled Fordson Power Major Tractor with power steering.
#175F. Cultivator, allocated #9536 in the 9000 Series.
#176F. Aerobat Rake, allocated #9537 in the 9000 Series.
#500F. Garage, wooden, allocated #9662 in the 9000 Series.
#503F. Thatched Cottage, allocated #9668 in the 9000 Series.
#505F. Enclosed Paddock.

Numbers not listed or illustrated were not used in this Series.

#1F. Model Farm display box art. *Courtesy Ray Haradin*.

#1F. Model Farm display. First issued in 1922 and containing 23 pieces. *Courtesy Ray Haradin*. $2,500-$3,000.

#4F. Tumbrel Cart with Carter in pre-War box. *Courtesy Brian Holmes*. $160-$180.

#4F. Tumbrel Cart with Carter in post-War box. Allocated #9505 in the 9000 Series and later issued with plastic horse. $130-$150.

#5F. Farm Wagon with Carter in pre-War box. $250-$300.

#5F. Farm Wagon with Carter in post-War box. This item was also issued both in a dark blue paint style with red trim and a cream paint style with red trim. See the *Great Book of Hollow Cast Figures* by Norman Joplin, New Cavendish Books. $220-$250.

#6F. General purpose plough. Two horses with Ploughman, post-War sliding box. $180-$220.

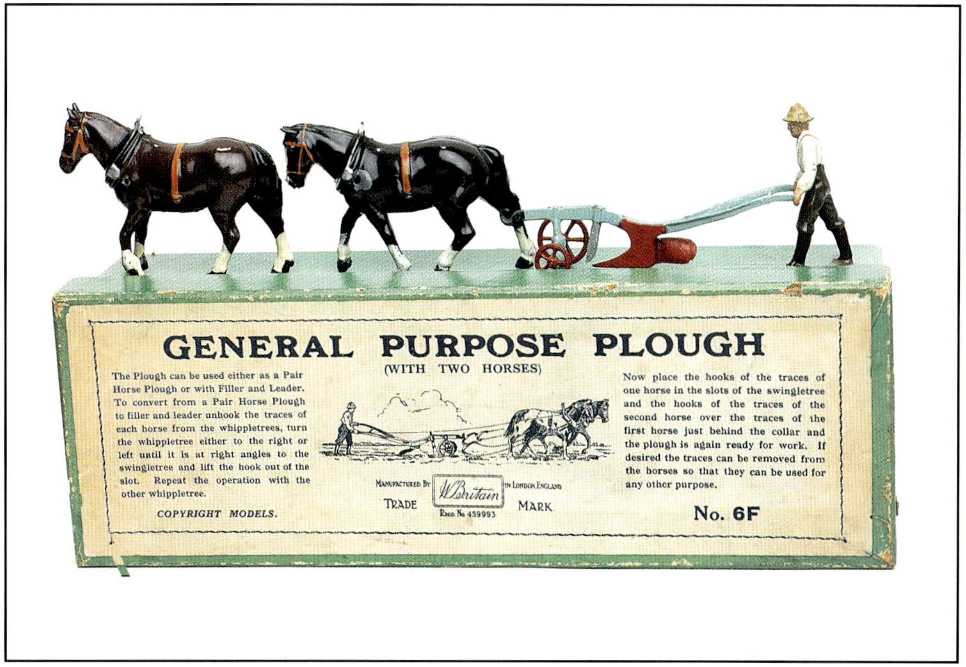

#6F. General purpose plough. Two horses with Ploughman, pre-War box. *Courtesy Brian Holmes*. $150-$250.

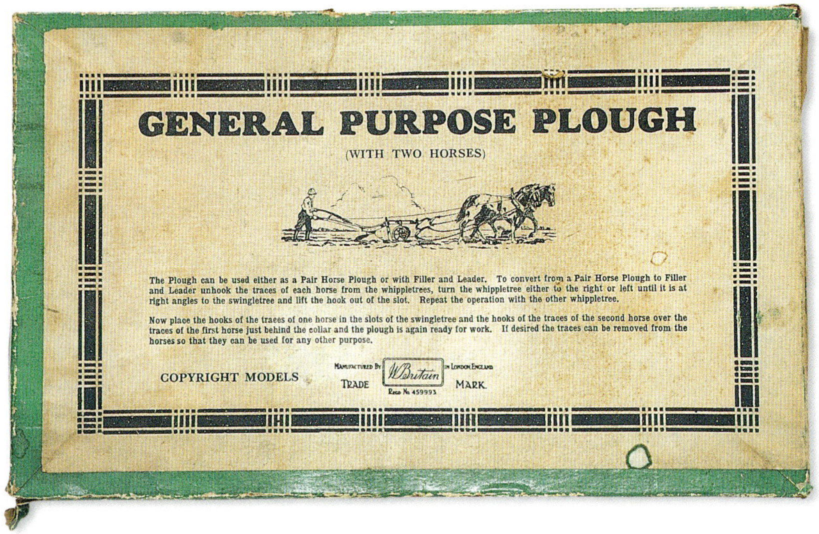

#6F. General purpose plough. Pre-War box art variation. $150-$250.

#7F. Tree and Gate. Pre-War box. $150-$250.

#7F. Tree and gate. Post-War box. *Courtesy Dudley Chignall*. $160-$220.

#8F. Horse rake and driver. Pre-War Box. *Courtesy Brian Holmes*. $100-$120.

#8F. Horse rake and driver. Post-War box. Allocated #9506 in 9000 Series. $100-$120.

#9F. Horse roller with Stable lad. Post-War box. Multi-section version roller. Allocated #9504. $100-$120.

#9F. Horse Roller with Stable lad, two-section version roller. $80-$100.

#9F. Horse Roller post-War box art variant. $100-$120.

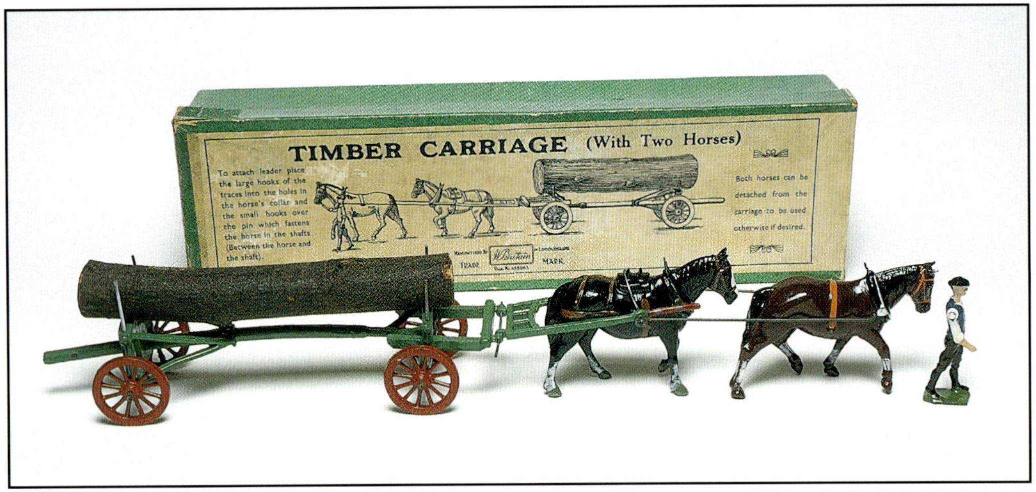

#12F. Timber Carriage with Stable lad. Pre-War box. *Courtesy Brian Holmes.* $300-$400.

#12F. Timber Carriage with Stable lad. Post-War square box version. $300-$350.

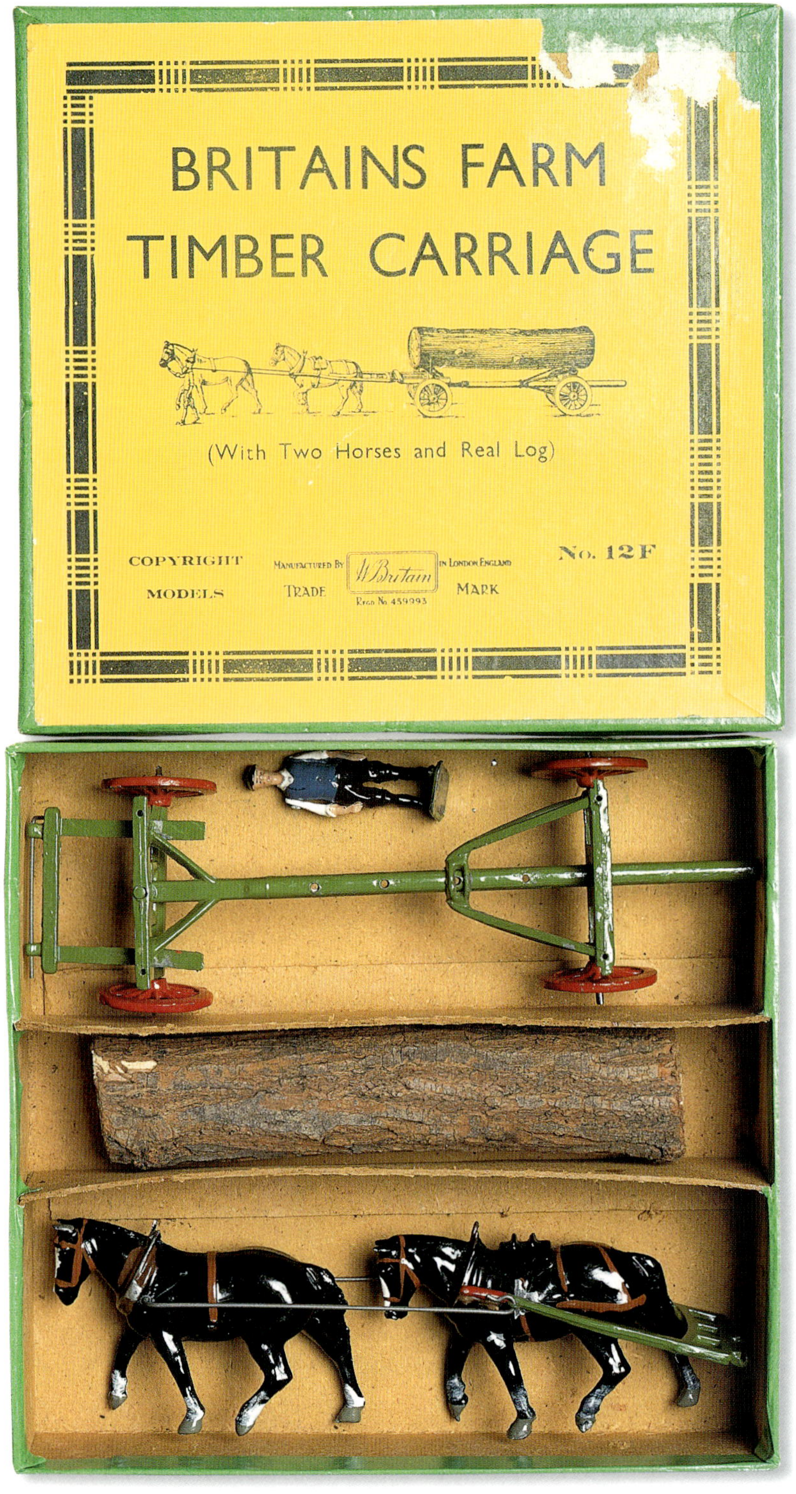

#12F. Timber Carriage with Stable lad. Square box version. See Farm Animal section for early fur hoof horse example. $350-$400.

#19F Tree and gate with boy on swing. Pre-War issue only. *Courtesy Brian Homes.* $180-$250.

#20F. Farmer's Gig with Farmer. Pre-War version, yellow gig and rare brown horse. Master Mold #161 was allocated to the Farmer and #141 to the Horse, which was not used in any other set. *Courtesy Philip Dean Reference Collection*. $280-$300.

#20F. Farmer's Gig with farmer. Pre-War box. $240-$260.

#20F. Farmer's Gig with farmer. Post-War box. *Courtesy Brian Holmes*. $220-$280.

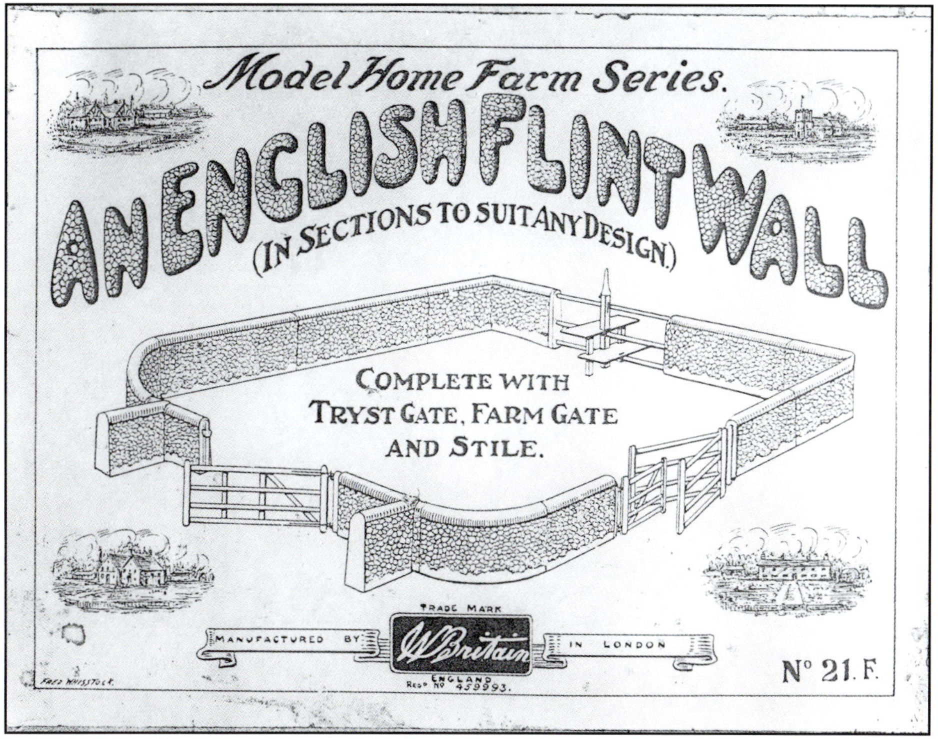

#21F. English Flint Wall box art. Pre-War issue only.

#22F. English Flint Wall display set. Pre-War issue only. *Courtesy Philip Dean Reference Collection*. $500-$600.

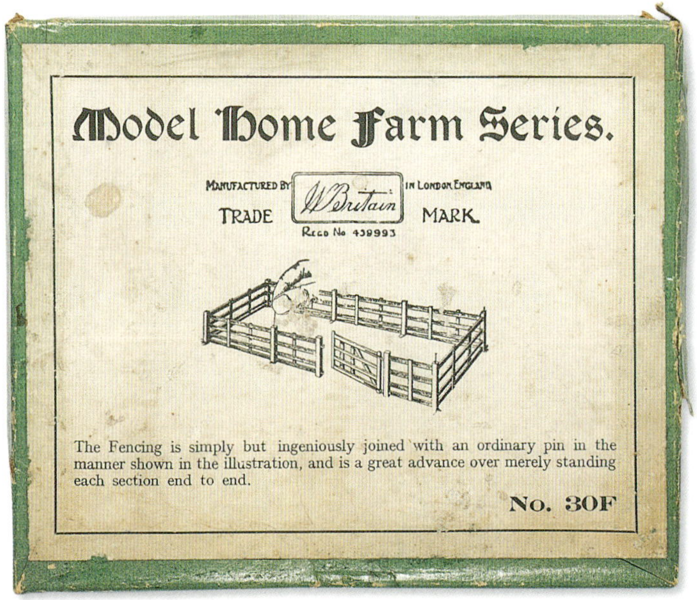

#30F. Green box art work.

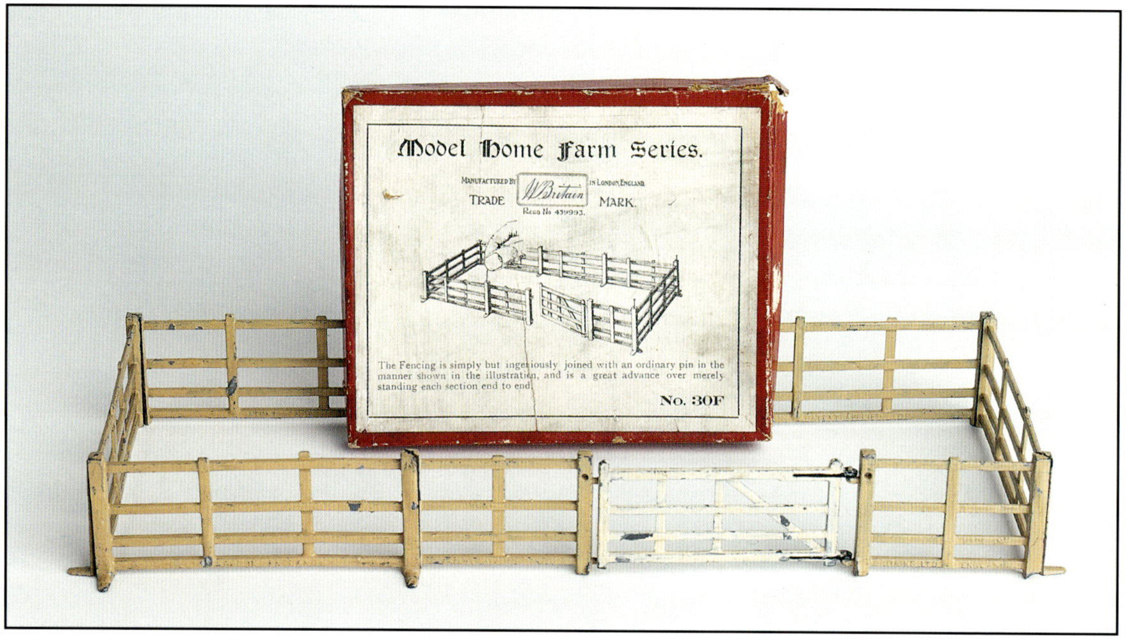

#30F. Fencing with gate. Pin type fence used only in #30F and #31F. Pre-War only. *Courtesy Philip Dean Reference Collection*. $400-$500.

#30F. Close up of unique pin type fence. *Courtesy Philip Dean*.

#31F. Fencing set. A larger box than #30F and a previously unrecorded set. *Courtesy Brian Holmes.*

#36F. Farm set showing red pre-War and green post-War box art. $120-$150 each.

#38F. Farm display. *Courtesy Christie's Auction House, London/The Arnold Rolak Reference Collection*. $180-$220.

#39F. Farm display. *Courtesy Bob Bailey*. $180-$220.

#40F. Farm cart. Pre-War box. The horse was allocated Master Mold #140. $80-$90.

#43F. Country Cottage with villagers, trees and flowers. *Courtesy Christie's Auction House, London, October 1992*. $2,200-$2,500.

#40F. Farm cart. Green post-War color variant. Allocated #9500 in the 9000 Series.; $80-$90.

#44F. Country cottage, paper mache. *Courtesy Philip Dean Reference Collection*. $600-$700.

#45F. Milk float and horse. Line drawing version box. The horse was allocated Master Mold #140 (see #131F). Allocated #9501 in the 9000 Series. $120-$150.

#53F. Home farm set. Post-War box. $120-$150.

#54F. Home farm set. Post-War box. $100-$130.

#55F. Home farm set. Post-War box. *Courtesy Bob Bailey.* $220-$250.

#56F. Home farm set. Post-War box.
Courtesy Bob Bailey. $140-$160.

#59F. Four wheel tipping lorry. Pre-War issue. $140-$160.

#58F. Fully Modelled Tree. Post-War box. A multi-sectional tree with branches that fit over the tree trunk. $120-$140.

#59F. Four wheel tipping lorry. Yellow version sometimes contained decals as a builder's lorry. See Road Chapter for a further example allocated #90F if issued as builder's lorry with decals. Pre-War only. *Courtesy Christie's Auction House, London/The Arnold Rolak Reference Collection.* $400-$450.

#59F. Four wheel tipping lorry. Post-War version showing tipping mechanism. $240-$260.

#60F. Six wheel lorry. Pre-War box. *Courtesy Christie's Auction House, London/The Arnold Rolak Reference Collection*. $380-$420.

#60F. Six wheel lorry, yellow paint variant. Pre-War only. Allocated #91F if issued with decals as a builder's lorry. *Courtesy Christie's Auction House, London/The Arnold Rolak Reference Collection*. $400-$500.

#60F. Six wheel lorry, rare blue version with red hub wheels. Pre-War only. $500-$600.

#61F. Ten wheel lorry. Pre-War only. $450-$550.

#61F. Ten wheel lorry, rare pale blue color variant with solid metal tires. Pre-War only. Allocated #92F if issued with decals as a builder's lorry. *Courtesy Christie's Auction House, London/The Arnold Rolak Reference Collection*. $600-$650.

#62F. Home farm display set. Post-War. *Courtesy Bob Bailey*. $250-$280.

#66F. Cardboard Farmhouse and scenic background display. Pre-War issue only (see #67F). *Courtesy Christie's Auction House, London/The Arnold Rolak Reference Collection*. $1,200-$1,500.

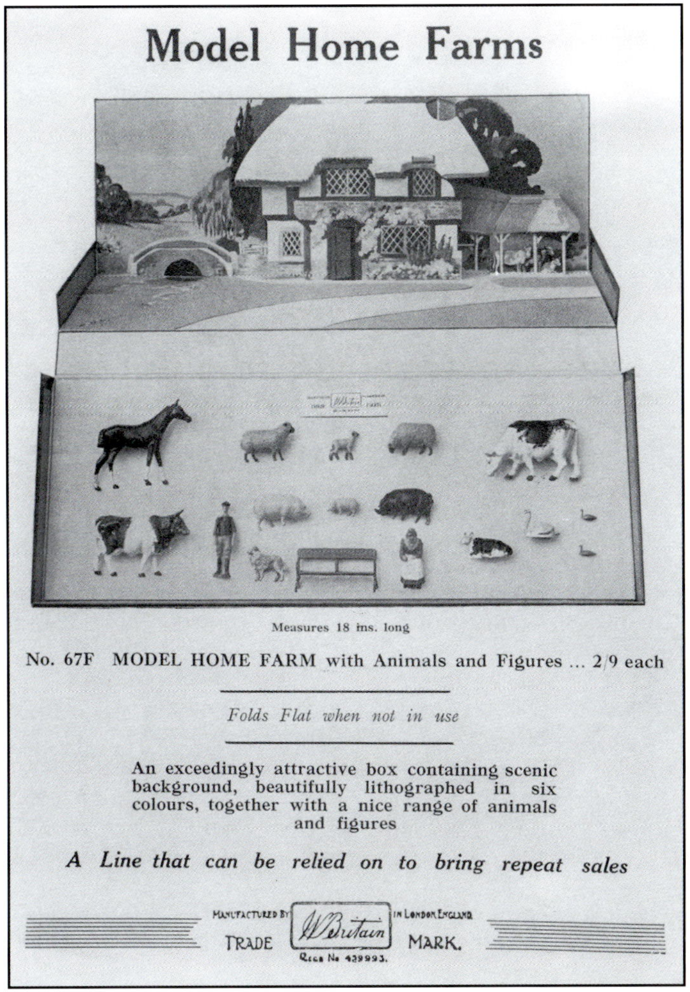
Model Home Farms

Measures 18 ins. long

No. 67F MODEL HOME FARM with Animals and Figures ... 2/9 each

Folds Flat when not in use

An exceedingly attractive box containing scenic background, beautifully lithographed in six colours, together with a nice range of animals and figures

A Line that can be relied on to bring repeat sales

MANUFACTURED BY W. Britain IN LONDON ENGLAND
TRADE MARK.
REGD No 439993.

#67F. A catalog page from 1935 advertising this rare set with fold out box and #66F cardboard backdrop.

#70F. Farm set. Pre-War only issue. *Courtesy Bob Bailey.* $250-$280.

#97F. Country Cottage, wooden. Pre-War only issue. $650-$700.

#94F. Farmhouse, wooden. Manufactured by Hugar. *Courtesy Philip Dean Reference Collection.* $600-$650.

#98F. Store Shed. Pre-War only issue. *Courtesy Christie's Auction House, London.* $300-$400.

#120F. Farm display set. Post-War only issue. *Courtesy Bob Bailey.* $180-$220.

#122F. Farm display set. Post-War only issue. *Courtesy Bob Bailey.* $180-$220.

#100F Pigsty, #101F Rabbit Hutch, and #102F Chicken Run. Pre-War issue only. $150 each.

#123F. Farm display set. Post-War only issue. *Courtesy Bob Bailey.* $100-$120.

#124F. Sheep and Sheepdog set. Post-War issue only. *Courtesy Bob Bailey.* $100-$120.

#125F. Cows and Calves set. Post-War issue only. *Courtesy Bob Bailey.* $80-$120.

#126F. Horse drawn farm cart with rubber tires (see also #40F). The horse was allocated Master Mold #140. Post-War issue only. Allocated #9502 in the 9000 Series. $100-$120.

#127F. Metal-wheeled Fordson Major Tractor with driver. Post-War issue only. *Courtesy Brian Holmes*. $300-$350.

#129F. Timber trailer with log, has towing hook to attach to #127F and #128F. Post-War issue only. Allocated #9560 in 9000 Series. *Courtesy Brian Holmes*. $120-$150.

#128F. Rubber-tired Fordson Major Tractor with driver. Post-War issue only. $300-$350.

#130F. Trailer with racks. Allocated #9550 in 9000 Series. Post-War issue only. $80-$100.

#133F. Farmyard display. Post-War only issue. $120-$130.

#130F. Trailer with racks. Line drawn box label variant. *Courtesy Christie's Auction House, London/ The Arnold Rolak Reference Collection*. $80-$100.

#134F. Tractors and implements display set. Post-War only issue. *Courtesy Bob Bailey*. $400-$450.

#131F. Horse-drawn milk float, milkman, and two milk churns. This is an improved version of #45F with milkman and churns added. Britains Dairy decals were added to this version of the milk float. The horse was allocated Master Mold #140. Post-War issue only. Allocated #9503 in the 9000 Series. $150-$180.

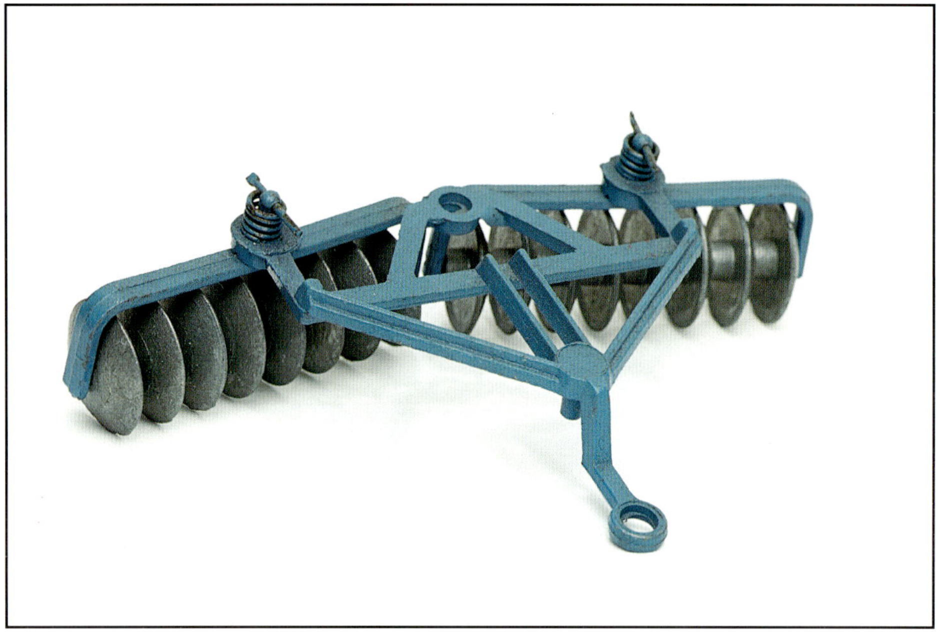

#135F. Disc Harrow. Post-War only issue. Allocated #9534 in the 9000 Series. *Courtesy Brian Holmes*. $50-$60.

#136F. Roller. Post-War only issue. Allocated #9533 in the 9000 Series. $50-$60.

#137F. Clockwork set with tractor, trailer and mechanical clockwork trailer. Post-War only issue. *Courtesy Brian Holmes*. $700-$800.

#138F. Four furrow tractor plough. Post-War only issue in sliding display box. Allocated #9551 in the 9000 Series. $50-$60.

#142F. Single horse general purpose plough and Ploughman. Post-War only issue. $140-$160.

#150F. Farmyard display. Illustrated with #152F. Post-War issues only. *Courtesy Christie's Auction House, London/The Arnold Rolak Reference Collection.* $200-$250.

#144F. Haystack, paper mache. Post-War only issue. $150-$250.

#147F. Farmyard display. Post-War issue only. *Courtesy Bob Bailey.* $80-$100.

#152F. Farmyard display. Post-War issue only. *Courtesy Bob Bailey*. $200-$250.

#155F. Farmyard display. Post-War issue only. *Courtesy Bob Bailey*. $140-$160.

#173F. Three furrow plough. Post-War only issue. Allocated #9530 in 9000 Series. *Courtesy Brian Holmes*. $50-$60.

#156F. Farmyard display. Post-War issue only. *Courtesy Bob Bailey*. $100-$120.

#174F. Muledozer. Post-War only issue. Allocated #9535 in 9000 Series. *Courtesy Brian Holmes*. $50-$60.

#501F. Cowshed. Post-War only issue. Allocated #9660 in 9000 Series and #502F Thatched Barn. Post-War only issue. Allocated #9663 in 9000 Series. *Courtesy Christie's Auction House, London, December 1997.* $600-$800 each.

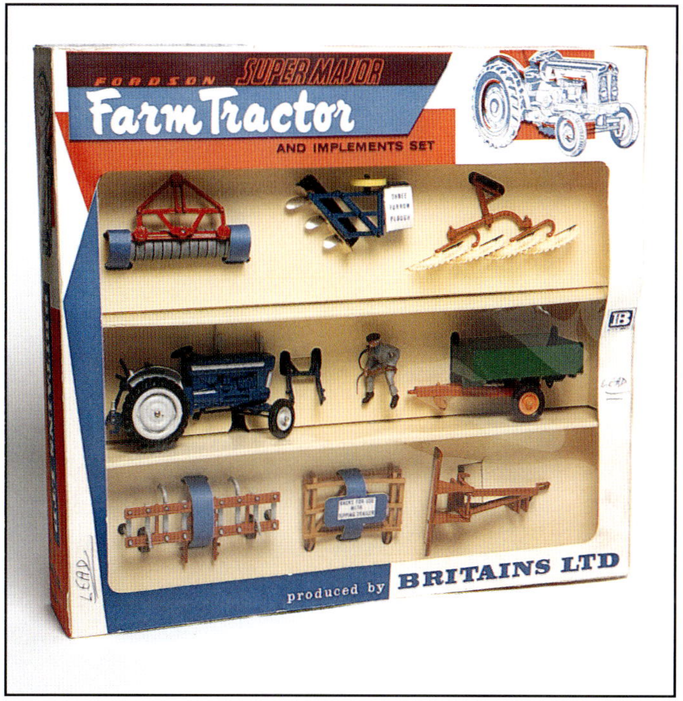

#9597. Rare 1962-65 isue from the Tractor and Implements Set. *Courtesy Bob Bailey.* $600-$650.

#504F. Large thatched farm house. Post-War issue only. Allocated #9668 in 9000 Series. *Courtesy Ray Haradin.* $800-$1,000.

#2041. Clockwork trailer. Post-War only issue. Also possibly used in Road Series with Britains Delivery Van without clockwork motor. $100-$120.

#2041. Clockwork trailer, green color variant. This item was originally issued as a khaki military trailer against this catalog number. $120-$150.

15. Farm Picture Packs, post-War

Farm Picture Pack boxes were issued in 1948 with combinations of Farm figures, Animals and Accessories from the Farm Range. Prices $35-$65 depending on number of components per box.

Farm Picture Pack Catalog issued by American Distributors Reeves International, Inc.

Farm Picture Pack box art.

Farm Picture Pack box art.

Farm Picture Pack box art.

Farm Picture Pack box art.

#5001 Shire Horse.

#5002 Cart Horse.

#5003 Field Horse.

#5004 Bull.

#5007 Cob and Foal.

#5005 Cow lying, Sheep with Lamb, and Exmoor Horn Sheep.

#5008 Milkmaid seated with milking pail, Cow, and Calf.

#5006 Shepherd, Sheep with Lamb, and Sheep feeding and walking.

#5009 Sheep walking and feeding, two Lamb, and six hurdles.

#5010 Farmer, Farmer's Wife, Goose, Gander, and Pig.

#5011 Six Hurdles, small feeding trough, two Pigs, and two Piglets.

#5012 Milkmaid carrying pail at side, Jersey Cow, and Calf lying.

#5013 Six field hurdles.

#5014 Highland Cattle.

#5015 Kennel with seated Dog.

#5016 Farmer, walking Sheep, Pi, and Collie Dog.

#5017 Cow lying and two calves lying.

#5018 Shepherd with crook, Collie Dog, Sheep walking and feeding, and two Lambs.

#5019 Cow walking.

#5020 Jersey Cow without bell.

#5021 Cow feeding.

#5022 Man with Wheelbarrow.

#5023 Horse feeding and small feeding trough.

#5024 Pig, two Piglets, and a Berkshire Pig.

#5025 Berkshire Pig.

#5026 Aged Villagers, Man and Woman seated on log seat.

#5027 Calf, Lamb, and two Goslings.

#5028 Farmer and Collie Dog.

#5031 Sheep walking and Lamb.

#5029 Farmer's Wife and three Hens.

#5032 Man with garden roller.

#5030 Pig, two Piglets, and a small feeding trough.

#5033 Cocks, Hens, Ducks, and Drakes; six pieces.

#5034 Man with swing water barrow.

#5037 Girl with feeding bucket and feeding Cock and Hen.

#5035 Collie Dog; Hounds standing, running, and head down; lying and seated dogs.

#5038 Huntsman with standing Hound.

#5036 Blacksmith, Anvil, and Cob.

#5039 Huntswoman with standing Hound.

16. Hunt, 1924

#608. Huntsman mounted, walking horse issued Pre- and post-War. Issued pre-War in set #1235 and described as Master Huntsman. A version wearing a top hat rather than a jockey cap appeared in set #1240. $55-$65 each. White Horse variant, $75-$85.

#598. Gentleman Farmer mounted, issued pre-War only in the Hunt Series, but also post-War in the Farm Range. $65-$75 each.

The first Hunt Series figures were introduced in 1924. Some of the single figures appeared in the numbering system for the Farm Series, while others were given catalog numbers within the main set numbering system.

In addition to the following illustrated items, the following boxed sets were issued, each containing at least one example of mounted or foot hunt figures. Set #234 The Meet, pre-War issue revived post-War and later adapted to the 9000 cellophane fronted box series (see illustration). Set #235 "Full Cry" as previous description (see illustration). Set #243 Huntsmen and Huntswomen with hounds, 7 piece set (not illustrated), pre-War only. Sets #1261 and #1262 were pre-War uncataloged sets. Each contained foot Hunt figures and hounds (not illustrated). Set #1445 Huntsmen and Huntswomen mounted and foot with fox and hounds (not illustrated).

#609. Huntswoman mounted sidesaddle, standing horse. Wears either bowler hat or top hat issued Pre- and post-War. Shown here with #614 standing hound. The Huntswoman also appeared in set #1238 pre-War only with either one or two hounds. Left to right: $55-$65, $25-$30, $75-$80.

#609. Huntswoman mounted sidesaddle, standing horse. Two rare paint variants both with bowler hats. The figure in the red hunting jacket was available only in the USA; the white/grey horse is a rare paint variant. Both examples are of pre-War vintage. $75-$80 each.

#609 Huntswoman mounted sidesaddle, standing horse. Rare red hunting jacket, pre-War issue available only in the U.S.A. with export tag. *Courtesy Thomas & Cathleen Wiedersiem*. $75-$80.

#610. Huntsman mounted on galloping horse, wearing either top hat, bowler hat, or jockey cap. Issued Pre- and post-War. Set #1237 contained this item with running hounds, pre-War, white horse. *Courtesy Brian Holmes*. $55-$65 each. White Horse variant, $75-$85.

#611. Huntswoman mounted sidesaddle on galloping horse wearing either top hat or bowler hat. Issued Pre- and post-War. Set #1238 contained this item with running hounds pre-War. $60-$65 each.

#613. Dismounted Huntsman, #614 Dismounted Huntswoman. The red coated figure on the right was issued only in the USA pre-War. The others were issued both Pre- and post-War. Left to right: $35-$45, $35-$45, $35-$45, $65-$75.

#614. Hounds standing. Two positions of Fox Hound issued against this catalog number. The Master Model records show that the Hound on the left with head up was given #212 and the Hound with head down #211. #615 Hound running issued Master Model #213. All were issued Pre- and post-War. $25-$35 each.

#616. Fox running. Issued Pre-and post-War. $50-$55 each.

#623. Huntswoman mounted astride wearing either bowler or rare top hat. Pre- and post-War. Set #1239 contained this item with standing hounds, pre-War. Left to right: $65-$70, $60-$65, $60-$65.

#1241. Gentleman farmer with hound. Pre-War set only. Sets #1235, #1237, #1238, and #1239 were combinations of single Hunt figures and hounds (not illustrated). See single item numbers 608, 609, 610 and 623. Left: $65-$75; right: $25-$35.

#1236. Huntswoman #609 with top hat. Rare pre-War issue only in this set with standing hound. *Courtesy Christie's Auction House, London/The Arnold Rolak Reference Collection*. $550-$600.

Export tag examples. Left: #608; right: #610. *Courtesy Joseph Kunzelmann*. $75-$85 each.

#1240. Huntsman #608 with top hat. Rare pre-War issue only in this set and standing hound. Left to right: $80-$90, $25-$35.

#608. Huntsman with top hat and hound, box art. Pre-War issue. *Courtesy Vectis Auctions Limited*.

Set #236 Hunt Series presentation set, pre-War box. Combining set #234 The Meet and set #235 Full Cry. *Courtesy Dr. Rob Wilson*. $2500-$3000.

Set #236, post-War box art. *Courtesy Ray Haradin*.

Set #1446. The Meet. Post-War version box. The set was also issued pre-War and was later reissued post-War as 9000 Series set #9650 (see illustration). $180-$200.

Set #236 Hunt Series presentation set, post-War box. Combining set #234 The Meet and set #235 Full Cry resulting in a 38-item display set. Also issued pre-War. *Courtesy Ray Haradin*. $2,400-$2,600.

A Britains post-War salesman's sample box containing Hunt Series figures. *Courtesy Ray Haradin.* Unique.

Top row, left to right: set #9650 (formerly set #1446) The Meet; set #9651 (formerly set #1447) Full Cry. Bottom Row, left to right: set #9655 The Meet (formerly set #234); set #9656 Full Cry (formerly set #235). All are in cellophane fronted boxes as was the case with all 9000 Series numbered sets. *Courtesy Christie's Auction House, London/Britains Archive 1994.* Top Row: $150-$170 each. Bottom Row: $500-$550 each.

Set #1447. Full Cry, post-War version box. The set was also issued pre-War and was later reissued post-War as 9000 Series set #9651 (see illustration). $180-$200.

A very rare wooden Manor House, Stable and Courtyard produced for F.A.O. Schwartz in the U.S.A. employing Hunt Series figures. *Courtesy William Miele.* $2,800-$3,000.

17. Racing Colors, 1925

Issued in 1925, these racehorses and jockeys in famous owners' colors were larger scale than standard; the horses stood 90mm high. The racehorses and colors catalog numbering system was somewhat spasmodic. Master Model numbers were as follows: Jockey #432B, Jockey and horse #123B and Racehorse #433B.

Catalog numbers 237 and 1463 were used pre-War. At first each contained a number of horses and jockeys. Some of the items were then allocated an individual catalog number. Post-War an RC # was used. In 1926, set #237 was used for six different racehorses and colors which were later issued as individual items: H.M. the King, #1480; Lord Derby, #1481; Lord Astor, #1482; and Lord Rosebery, #1488. Neither Lord Londonderry or Lord Woolavington were reissued with an individual number. Major Noel Furlong, #1483; The Duke of Portland, #1485; and Sir Hugo Cunliffe-Owen, #1484 were added. Lord Londonderry and Lord Woolavington were deleted around 1936. Set #1463 was used for six individual figures in 1936 and as a combined box of six in 1938: His Highness the Aga Khan, #1486; Lord Glanely, #1487; Miss Dorothy Paget, #1489; Mrs. J. de Selincourt, #1490; Lord Mildmay of Flete, #1491; and Lord Rosebery, #1488 (previously also part of set #237).

British Racing Colors not illustrated.

#RC1. H.M. The King, pre- and post-War. Replaced by H.M. The Queen. Pre-War H.M. The King was part of set #237 with an individual catalog number of 1480.

Lord Londonderry, pre-War, part of set #237 not allocated an individual catalog number.

Duke of Portland, pre-War, part of set #237 allocated individual #1485.

Date of issue in brackets. Prices: A thirty percent premium should be added when boxed.

American Racing Colors: *Indicates Not Illustrated

#RC15. Glen Riddle Farm
#RC16. Greentree Stable #1830
#RC24. Alfred Vanderbilt #1714
#RC93. J.H. Whitney #1830
#RC94. C.V. Whitney #1829
#RC99. Calumet Farm
#RC119. Milky Way Farm, Mrs. E.V. Mars
#RC124. Wheatley Stables #1829*
#RC127. Charles S. Howard
#RC130. Brookemeade Stables*
#RC136. Joseph E. Widener #1829*
#RC162. King Ranch*
#RC163. Gain Hoy Stables*
Mr. T.D. Buhl #1713*
Mr. Marshall Field #1829*
Mr. Howard W. Maxwell #1829*
Mr. J.H. Loucheim #1829*
Mr. Edward R. Bradley #1830*
Bing Crosby Stables #1830*
Mrs. E. Denemark #1830*
August Belmont #1830*

All Racing Colors with an RC# and a four digit number were issued both pre- and post-War. The others are pre-War only items.

Six colors were grouped under #1829 and #1830 pre-War.

Canadian Racing Colors, post-War. *Indicates not illustrated individually, but in display set illustration.

#RC154. Mr. W. R. Beasley
#RC155. Maloney and Smythe*
#RC156. Stafford Farms
#RC157. Lanson Farm* The salesman's sample box containing this item uses the spelling "Lansom," M.J. Boylen.
#RC158. E.P. Taylor
#RC159. Seagram Stables
#RC160. Mr. R. James Speers
#RC161. Mr. R.S. McLaughlin*

Extensive research has been carried out by British Collector Peter Kirk. All of the information and illustrations in this chapter was supplied by Peter and permission granted to use the information from Peter's book *Racing Colors of Famous Owners*.

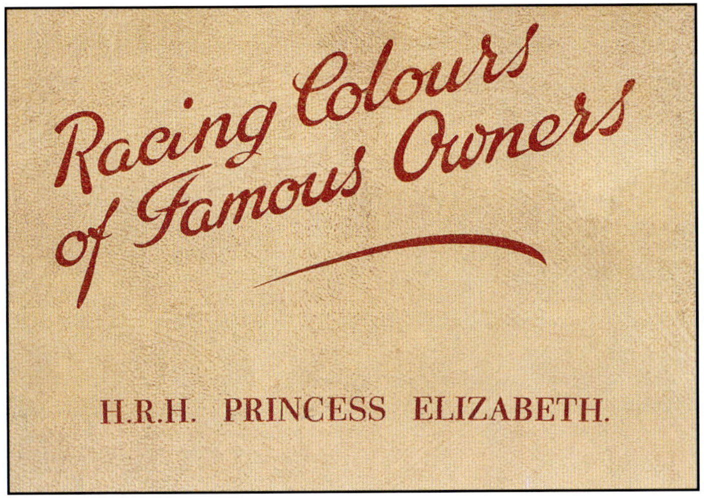

Racing Colors of Famous Owners, post-War box art.
Courtesy Philip Dean.

H.M. The Queen in scenic box, #RC1 (1958). Pre-War #RC1 was #1480 and #237. His Majesty The King. Royal Colors remain the same (not illustrated). $180-$200. H.M. The King unboxed $100-$150; boxed $180-$300 (not illustrated).

Left to right: #RC14 H.M. Queen Elizabeth, The Queen Mother (1954); #RC1 H.M. The Queen (1958); and #RC82 H.R.H. Princess Elizabeth (1951). Left to right: $150-$180, $180-$200, $250-$300.

H.R.H. Princess Elizabeth #RC82 with original box (1951). $280-$350.

Left to right: Sir Victor Sassoon #RC8; Mr. J.V. Rank #RC12; and Glen Riddle Farm #RC15 (1954). Left to right: $120-$150, $120-$150, $180-$220.

Left to right: Lord Astor #RC2, pre-War #237 and #1482; Lord Derby #RC3, pre-War #237 and #1481; and Lord Rosebery #RC6 (1954), pre-War included in #237 and #1463 given its own number 1488 in 1937. $120-$150 each.

Left to right: Greentree Stable #RC16; Alfred Vanderbilt #RC24; and Prince Aly Khan #RC31 (1954). Left to right: $180-$220, $180-$220, $120-$180.

Opposite page;

Bottom left: Left to right: Calumet Farm #RC99; Milky Way Farm #RC119; and Charles S. Howard #RC127 (1954). $180-$220 each.

Bottom right: Left to right: Mr. J.E. Ferguson #RC150; Lord Gainsborough #RC151; and Mr. J. Olding #RC152 (1954). $120-$180 each.

Left to right: H.H. The Aga Khan #RC63, pre-War in set #1463 and individual #1486; Miss Dorothy Paget #RC64, pre-War set #1463, individual #1489; Mr. H.J. Joel #RC67. $120-$180 each.

Left to right: M. Boussac #RC144; Lord Milford #RC145; and M.M. Leon Volterra #RC146 (1954). $120-$180 each.

Left to right: Mr. Winston Churchill #RC83; J.H. Whitney #RC93; and C.V. Whitney #RC94 (1954). Left to right: $120-$180, $180-$220, $180-$220.

Left to right: Mr. J.L. Jarvis #RC147; H.H.M. Setta Devi, The Gaekwar of Baroda #RC148; and Baron Guy de Rothschild #RC149 (1954). $120-$180 each.

Left to right: Duke of Norfolk #RC142; Mr. S. Wootton #RC153; and Mr. W. R. Beasley #RC154 (1958). Left to right: $120-$180, $120-$180, $180-$220.

Left to right: Mr. S. Wootton #RC153; Mr. W. R. Beasley #RC154; and Mr. E.P. Taylor #RC158 (1958). Left to right: $120-$180, $180-$220, $180-$220.

Left to right: Stafford Farms #RC156; Seagram Stables #RC159; and Mr. R. James Speers #RC160. $180-$220 each.

Left to right: Sir Hugo Cunliffe Owen #1484 (1936); Lord Woolavington #237 (1926) and Sir John Jarvis. $120-$180 each.

Left to right: Duke of Norfolk (1954); #RC142 Duchess of Norfolk, Mickey Mouse colors; Duchess of Norfolk registered colors (1938). Left to right: $120-$180, $300-$400, $120-$180.

Left to right: Lord Glanely set #1463 and individual #1487; Lord Mildmay of Flete (first color, 1937) #1491; and Mrs. J. de Selincourt (1936) #1490. All issued as catalog #1463. $120-$180 each.

Left to right: Lord Rosebery catalog #1436 (1936); Lord Rosebery catalog #237 (1926) with original pre-War boxes. Slightly different paint styles. $250-$300 each.

Left to right: Lord Ailesbury (1947); Lady Ashcombe (1953); and Mr. P.T. Beasley (1948). $120-$180 each.

Left to right: Lord Bicester (1941); Sir Alfred Butt (1936); and Mr. W.A. Chanler (1928). $120-$180 each.

Left to right: Captain C.F. Elsey (1927); Foxcatcher Farm (1937); and Major Noel Furlong (1936) #1483 and in set #237. $120-$180 each.

Left to right: Mr. G. W. Hands (1936); Mr. Owen Howells; and Lord Latimer (1941). $120-$180 each.

Left to right: Lord Milton (1936); Captain Lionel Montagu (1934); and Mrs. McDonald-Buchanan (1938). $120-$180 each.

Left to right: Mr. F.S. Myersgough (1928); Mr. Henry Roberts; and Mr. R. Rounds (1938). $120-$180 each.

Left to right: Mrs. Scott (1938); unidentified owner's color; and Marquis of Zetland (1930). $120-$180 each.

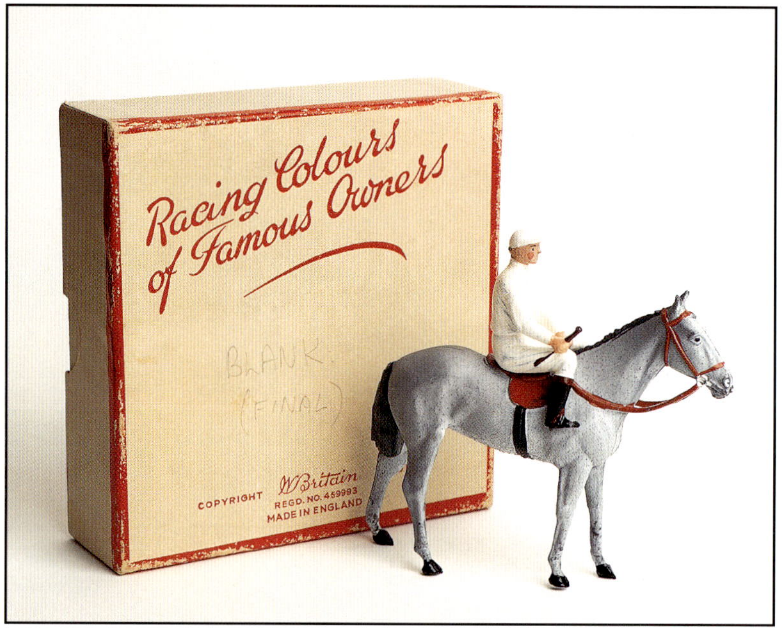

Plain white to enable the collector to paint in their chosen color catalog #RC0 (1954), pre-War #1493. $220-$250.

Left to right: Mr. J.A.D. de Rothschild (1937) pre-War in set #1463, individual #1492; and Baron Guy de Rothschild (1954). Family comparison colors. $120-$180 each.

Many famous owners racing colors remain unidentified. An example is shown here. See *The Christie's Britains Archive Catalog 1994* for further examples. Peter Kirk's research has corrected some which were misidentified in the auction and details appear in Peter's Publication *Racing Colors of Famous Owners*. Unique.

Examples of pre-War, post-War, and post-War scenic boxes.

Duchess of Norfolk Mickey Mouse colors with original box issued by Britains in 1938 as catalog #1551. $800-$1,000.

Salesman's sample. Canadian Racing Colors. *Courtesy New Cavendish Books/British Tin Toys, Marguerite Fawdry*. Unique.

Duchess of Norfolk close up of Mickey Mouse on jockey's shirt.

18. Police and Road, 1927

Automobile Association, Police, St. John's Ambulance and Air Raid.

Automobile Association men and signs were the first items to be produced connected with road, traffic, or civilian vehicles, when, in1927, of set #239 appeared. Apart from the standing policeman in helmet included in the Railway Range in 1909, the first police items commenced in 1929 with set #319. Civilian vehicles and sports racing cars appeared first in 1935 with St. John's Ambulance figures and ambulances being introduced in 1936 and 1937, respectively. A series of clockwork and non-clockwork delivery vehicles was manufactured post-War, the other items having been eliminated by 1941.

A rare R.A.C. (Royal Automobile Club) man, based on the A.A. figure. *Courtesy Ray Williams*. Unique.

Set #239 Automobile Association Motor Patrol and Signs also found in a box marked "Givjoy", an English company operated by a Mr. Halliday, producer of Skybird Aircraft. Givjoy was the company's trade name and were thought to have commissioned the set from Britains. All components of this set appear also in the Farm chapters. *Courtesy Bob Bailey.* $600-$800.

Set #319, Police mounted and foot. *Courtesy Bob Bailey*. $600-$800.

Set #1413, Police car with two occupants. Master Model #550B was allocated to the occupants and #972 to the car. *Courtesy Philip Dean Reference Collection*. $800-$900.

U.S. Policeman directing traffic. Single figure from rare set #428 issued only for one year, 1931-1932. $80-$90.

Set #1413, Police car, color variant. *Courtesy Ray Haradin*. $1,000-$1,200.

Set #1426, St. John's Ambulance Brigade. Items in this set were not given individual catalog numbers, therefore, the numbers given are the Master Model numbers. Stretcher Bearer #372B; Wounded Civilian #373B; Nurse #374B; Private (pictured left) #375B; Officer (pictured right) #376B. The Stretcher was given Master Mold #741. All these figures came painted in both gloss and matte, the latter at the request of the St. John's Ambulance Brigade. *Courtesy Bob Bailey.* NB The St. John's Ambulance Brigade is a voluntary organization. These figures, therefore, are compatible with the Volunteer Corps Ambulance illustrated in the Vehicle section. $1,600-$1,800.

A very rare St. John's Ambulance Stretcher Bearer with steel helmet. *Courtesy Dr. Rob Wilson.* Unique.

Road signs. Left to right: Steep Hill #622B; "Z" Bend #623B; Hospital #624B; Level Crossing #625B; Slow Major Road #626B; 30 MPH sign #627B. *Courtesy Philip Dean Reference Collection.* $45-$55 each.

Set #1428, Road Signs and Traffic Lights. *Courtesy Philip Dean.* Set numbers #1427 and #1429 contained various assortments of road signs. #1430 was a large traffic display with scenic background and contained various items from the Farm and Garden Series (not illustrated). A rare single row set of road signs is illustrated in *The Great Book of Hollow Cast* Figures by Norman Joplin, Published by New Cavendish Books and is incorrectly attributed to set #1428. $600-$700.

Road signs. Left to right: No Speed Limit (reverse side of 30 MPH sign) #627B; Crossroads #628B; "T" Junction #629B; Narrow Bridge #630B; Crossing, No Gates, #631B; Road Junction #632B; Halt Major Road #633B. *Courtesy Philip Dean Reference Collection*. \$45-\$55 each.

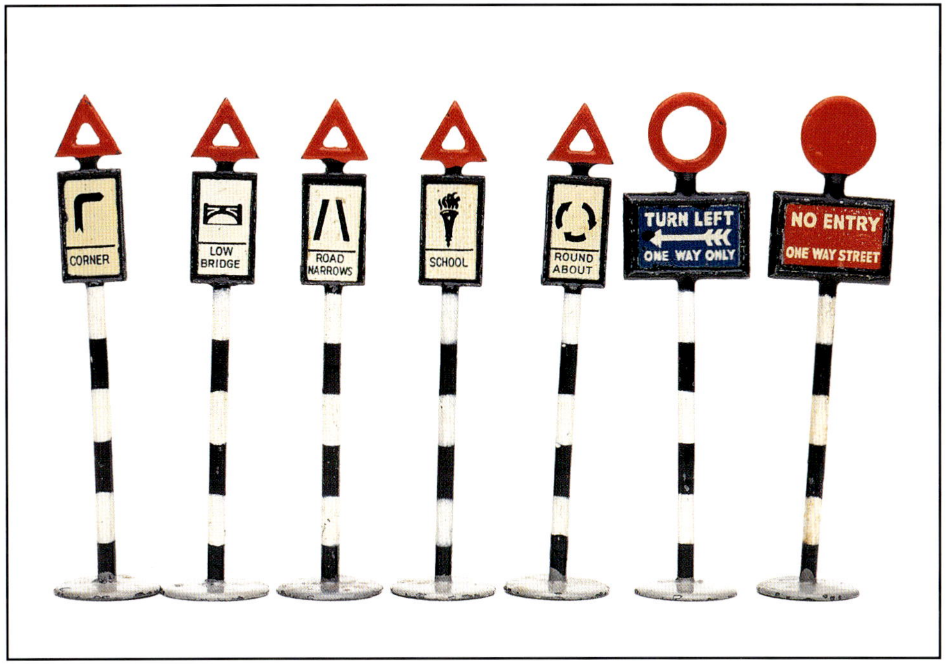

Road signs. Left to right: Corner #634B; Low Bridge #635B; Road Narrows #636B; School #637B; Roundabout #638B; Turn Left #639B; No Entry #640B. *Courtesy Philip Dean Reference Collection*. \$45-\$55 each.

Kiddicraft Traffic set box art. This box contained one of each of all the Britains Road Signs, along with a number of Civilian figures and two Minic Tinplate Vehicles. *Courtesy Philip Dean Reference Collection*. \$600-\$700.

Street Furniture. Belisha Beacon, #348B; Lamp Post, #349B; Traffic Lights, two sets, #350B. Numbers are from the Master Model list. \$45-\$55 each.

Set #1511 Mounted Policemen, also used in #1477 Coronation display set. Allocated Master Model #421B. *Courtesy Bob Bailey.* $350-$400.

Set #1792 Traffic Police Motorcycles. Master Model #528B was allocated to the second grade item illustrated on the left and #593B to the two first grade items. Left to right: $55-$65, $80-$90, $80-$90.

Air Raid Warden Master Model #1106B. *Courtesy Christie's Auction House, London.* $200-$240.

Set #1914 Air Raid Wardens. *Courtesy Bob Bailey.* $3,000-$3,500.

Set #1868 Air Raid Warden's Post, wooden, with gas detector platform. *Courtesy Christie's Auction House, London.* $2,200-$2,500.

Set #90F, rare Davis Estates, Ltd. Builder's Lorry. This four wheel version is Farm Lorry #59F with added decals; #91F, six wheel and #92F, ten wheel, are also adapted from the corresponding Farm Lorry set numbers. *Courtesy Christie's Auction House, London.* $3,500-$4,000.

Left: set #1393 Bluebird Speed Record Car, a one-piece casting. See illustration of set #1400 for comparison. Right: #329B, a smaller version Bluebird made for Cadbury as part of a gift scheme for members of the Cococub Club (see Cococub chapter). *Courtesy Christie's Auction House, London/The Arnold Rolak Reference Collection.* $350-$400 and $400-$450.

Set #1392 Civilian Autogiro. *Courtesy Bob Bailey.* $2,500-$3,000.

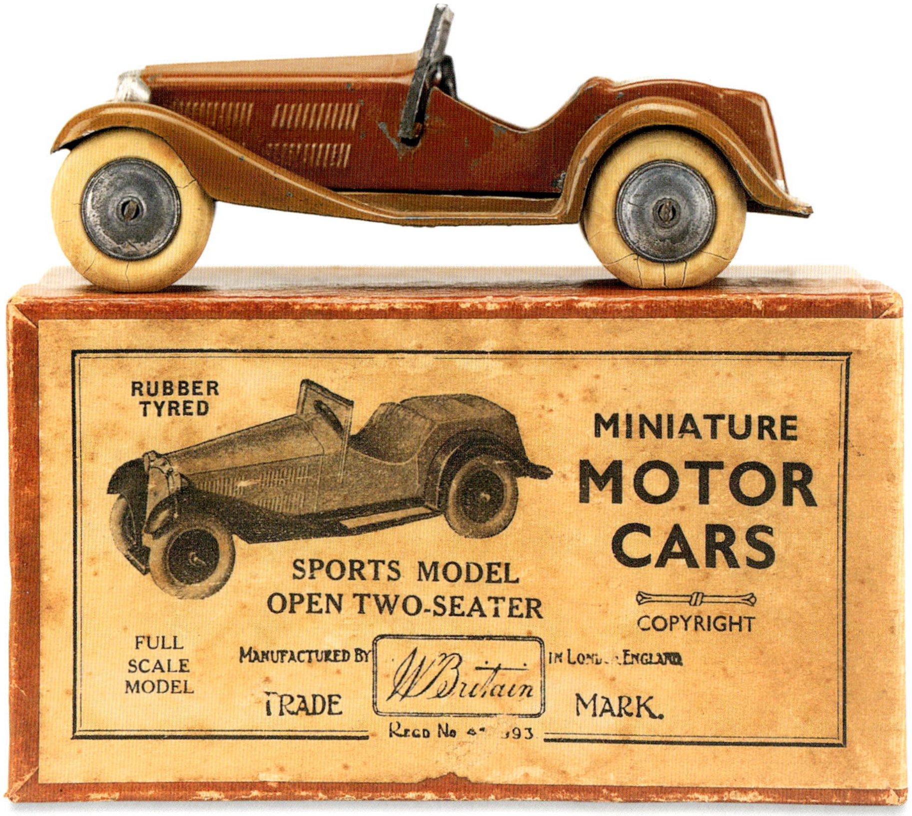

Set #1398 Sports Model Open Tourer Motor Car. *Courtesy Christie's Auction House, London/The Arnold Rolak Reference Collection.* $1,500-$1,800.

Set #1399 Two-Seater Coupe Model Motor Car in rare cream and black version, with box which has original one shilling price sticker. *Courtesy Christie's Auction House, London June 24, 1999.* $1,800-$2,000.

Set #1398 Sports Model Open Tourer Motor Car, rare cream and red version. *Courtesy Philip Dean Reference Collection.* $800-$900.

Set #1399 Two-Seater Coupe Model Motor Car. *Courtesy Ray Haradin.* $2,800-$3,000.

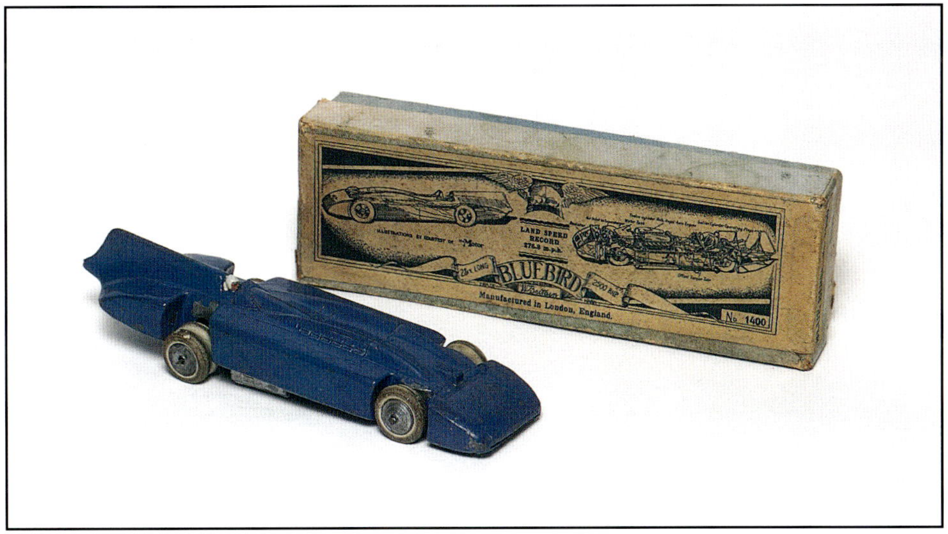

Set #1400 Bluebird Speed Record Car (see set #1393). *Courtesy Bob Bailey.* $350-$400.

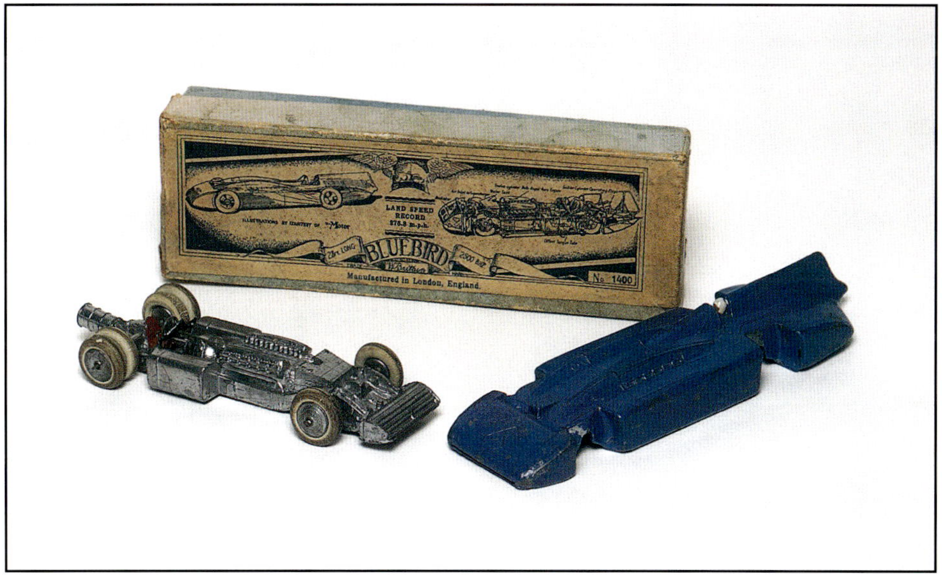

Set #1400 Bluebird Speed Record Car showing removable chassis (see set #1393). *Courtesy Bob Bailey.* $350-$400.

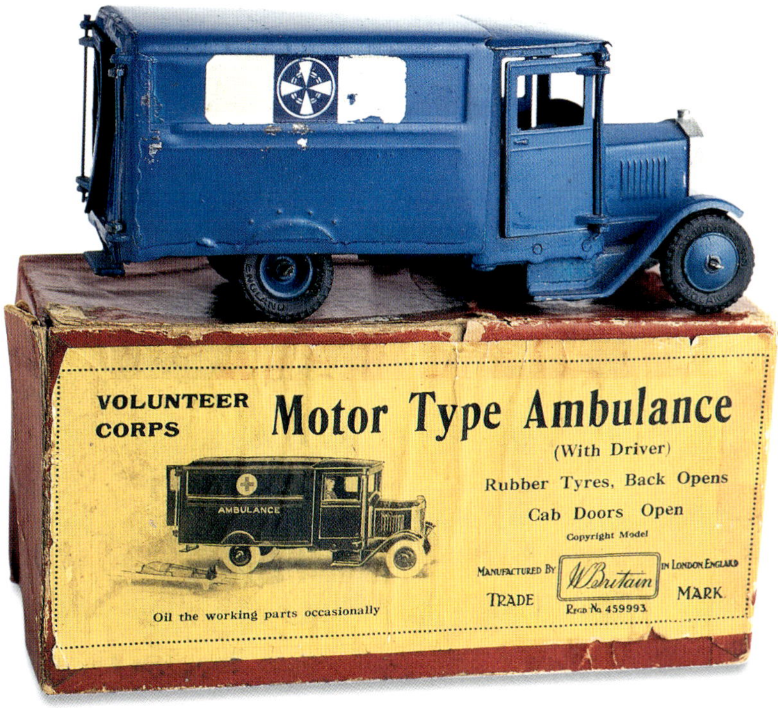

Set #1513 Volunteer Corps Motor Ambulance with driver, allocated Master Model #1195B, wounded, and stretcher. This vehicle has the St. John's Ambulance Brigade emblem on each side and would, therefore, compliment set #1426. *Courtesy Christie's Auction House, London/The Arnold Rolak Reference Collection.* $900-$1,200.

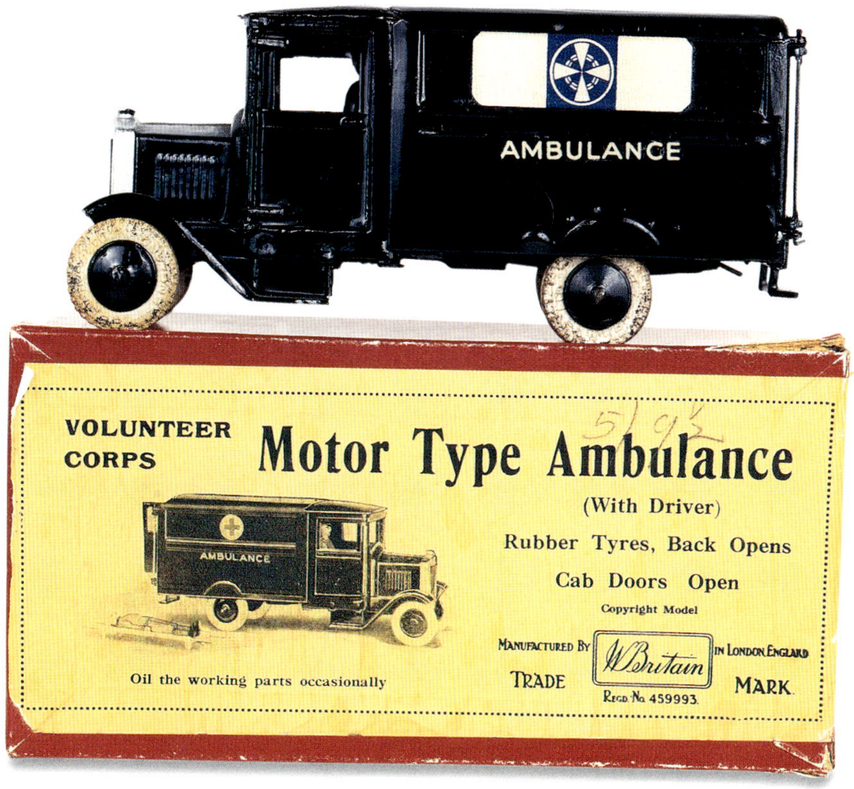

Set #1513 Volunteer Corps Motor Ambulance, dark blue paint variant. *Courtesy Philip Dean Reference Collection.* $900-$1,200.

Set #1513 Volunteer Corps Motor Ambulance, round cab, with driver, wounded, and stretcher. Post-War issue, mid blue. The St. John's Ambulance figure set #1426 was not reissued post-War. *Courtesy Christie's Auction House, London/The Arnold Rolak Reference Collection.* $600-$700.

Set #1513 Volunteer Corps Motor Ambulance, round cab, in original box showing driver, wounded, stretcher, and opening doors. *Courtesy Bob Bailey.* $900-$1,000.

Set #1513, Volunteer Corps Motor Ambulance, round cab, with driver, wounded, and stretcher. Post-War issue, dark blue paint variant. *Courtesy Christie's Auction House, London/The Arnold Rolak Reference Collection.* $600-$700.

Set #1513 Volunteer Corps Motor Ambulance, rare split windscreen. Post-War version with red hub wheels. *Courtesy Christie's Auction House, London/The Arnold Rolak Reference Collection.* $700-$800.

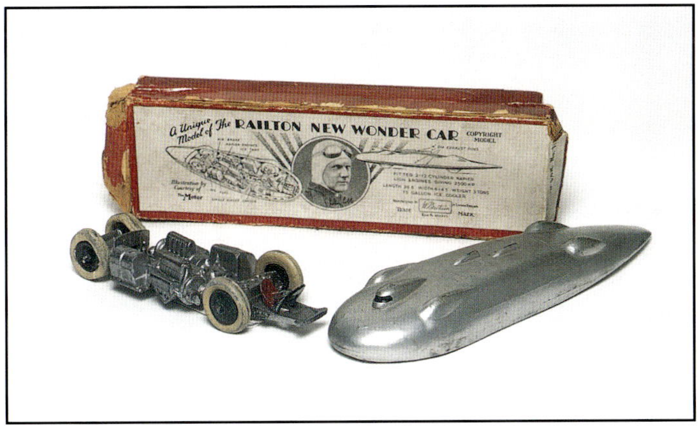

Set #1656 John Cobb's Railton Wonder Car. Chassis is detachable. *Courtesy Bob Bailey.* $400-$450.

Set #1552 Royal Mail Van with driver allocated Master Model #1004B. *Courtesy Ray Haradin.* $8,000-$9,000.

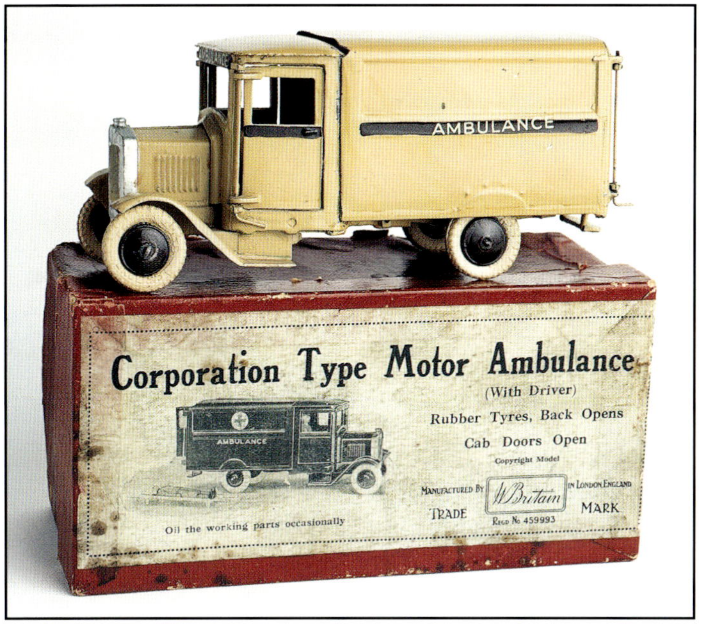

Set #1514 Corporation Type Motor Ambulance with driver allocated Master Model #985B, wounded, and stretcher. The cream color indicates that this was a hospital ambulance and not manned by the St. John's Ambulance Brigade. *Courtesy Christie's Auction House, London/The Arnold Rolak Reference Collection.* $1,000-$1,500.

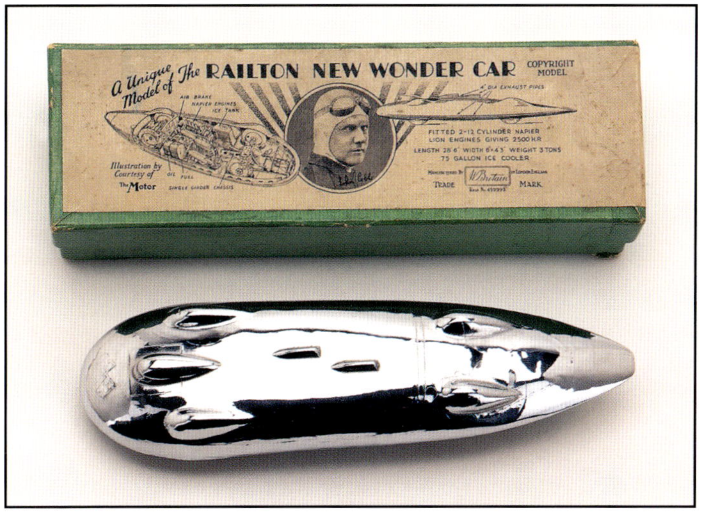

Set #1658 John Cobb's Railton Wonder Car, chromium plated version of #1656. *Courtesy Christie's Auction House, London, March 1997.* $450-$500.

Set #2024 Light Goods Van with "Britains Ltd." logo on side and original box. *Courtesy Philip Dean.* $600-$700.

Set #2045 Clockwork Van with "Britains Ltd." logo on side with original box and key. *Courtesy Bob Bailey.* $800-$900.

Set #2024 Light Goods Van with "Britains Ltd." logo on side. *Courtesy Christie's Auction House, London/The Arnold Rolak Reference Collection.* $500-$550.

Set #2045 Clockwork Van with "Britains Ltd." logo on side with original box. *Courtesy Christie's Auction House, London/The Arnold Rolak Reference Collection.* $800-$900.

Set #2045 Clockwork Van with "Britains Ltd." logo on side; rare blue and cream color variant. $800-$900.

A rare brown paint variant of #59F Farm Lorry. Driver allocated Master Model #1475B. *Courtesy Brian Holmes*. Unique.

Garage Series

#748 first version Shell petrol pump. This pump was changed to tan in color and a fan shaped logo added. It was reallocated Catalog #1V. See #101V. *Courtesy Philip Dean Reference Collection*. $40-$45.

Experimental colors that were not put into production. Left to right: #1398 Sports Model Open Tourer Motor Car and #1399 Two-Seater Coupe Model Motor Car, both with a post-War paint finish. A round cab six wheel lorry in red livery based on the pre-War #60F farm vehicle. However, this must have been a prototype as #60F was not issued post-War. *Courtesy Christie's Auction House, London/Britains Archive 1994*. Unique.

#749 B.P. petrol pump and #751 Power petrol pump. The B.P. petrol pump was later given a shield logo, changed to cream in color, and reallocated Catalog #2V. See #101V. $40-$45 each.

#748 Shell petrol pump; #749 B.P. petrol pump. *Courtesy Philip Dean Reference Collection.*

The following petrol pumps are not illustrated:

#771 Esso, yellow color variant
#7V Dominion.
No petrol pump was allocated #8V

Garagehands with petrol pump nozzle. These figures appeared in the catalog as #787. *Courtesy Philip Dean Reference Collection.* $80-$90 each.

#750 Shellmex petrol pump. *Courtesy Philip Dean Reference Collection.*

Top left: Garagehands/Petrol pump attendants #787. *Courtesy Stewart Saxe.* $80-$90 each.

Top right: Set #9689, previously #104V, Garagehand, Shell Oil cabinet, Shell, B.P., and National Benzole petrol pumps #9671, #9675, #9672 respectively on stand with original box. For catalog numbers for pumps see previous illustrations. *Courtesy Philip Dean Reference Collection.* $320-$350.

Right: #9V Castrol and #10V Shell oil cabinets. *Courtesy Christie's Auction House, London/The Arnold Rolak Reference Collection.* $40-$45.

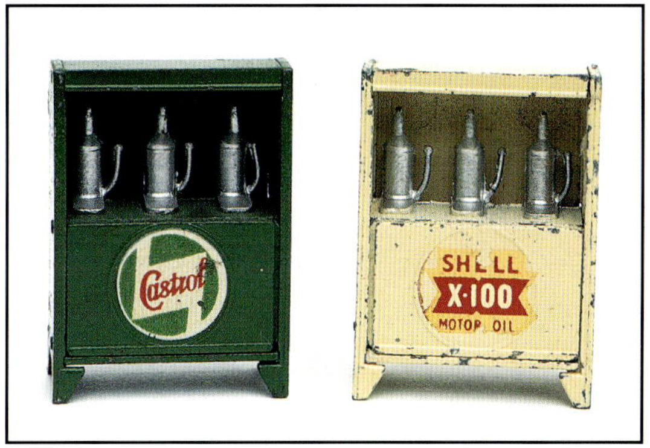

Set #102V, later #9682. Esso, Mobil Gas, and Fina petrol pumps with stand. #3V, #4V and #6V, respectively. *Courtesy Philip Dean Reference Collection.* $180-$220.

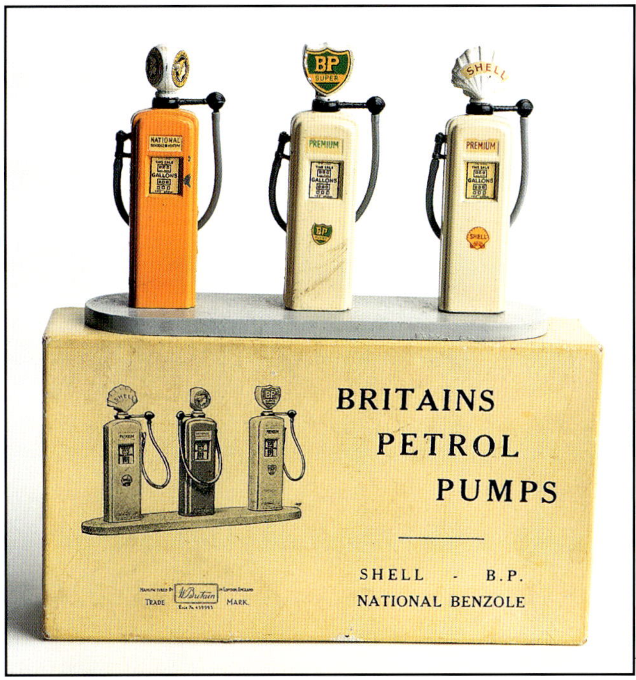

Set #101V Shell, B.P., and National Benzole petrol pumps on stand with original box. Catalog numbers for the pumps were #5V, #2V and #1V, respectively. *Courtesy Philip Dean Reference Collection.* $240-$260.

Set #103V, later #9683. Shell, Esso, and Mobil Gas petrol pumps with stand. #1V, #3V and #5V, respectively. *Courtesy Philip Dean Reference Collection.* $180-$220.

19. Zoo, 1930

The Zoo Series was issued first in 1930 and, with the addition of several new animals in 1953, remained in the catalogs until 1959/1960. The original inspiration for most of the figures came from London's Regent's Park Zoo where Dennis Britain and others from the company took sketches of the animals. As well as being featured in all of the catalogs between 1930 and 1960, a catalog devoted only to the Zoo range was produced sometime between 1930 and 1941. A foldout sheet in the back contains catalog numbers beyond those allocated descriptions and it is possible that this was in readiness for new designs to be allocated a number. The Roan Antelope illustrated in the following pages perhaps confirms this theory.

Pre-War Zoo sets not illustrated

#2Z. Comprised of Lion, Lioness, Camel, Zebra, Polar Bear sitting, Kangaroo, Gibbon Monkey climbing, Guenon Monkey, Pelican beak open, Pelican beak closed, two Penguins and two King Penguins. (14 pieces)

#5Z. Camel, Crocodile, Eland Bull, Polar Bear sitting, Ostrich, Llama, Bison, Brown Bear, Lion, Lioness, Kangaroo, Giraffe, Hippopotamus, Rhinoceros, Zebra, two Date Palms and two Coconut Palms. (19 pieces)

#6Z. Comprised of Brown Bear, Gibbon, Monkey climbing, Guenon Monkey, Chimpanzee, Eland Bull, Ostrich, two Bear Cubs, Bison and Llama. (10 pieces)

#17Z. Comprised of Giant Tortoise, young Crocodile, Stork, Polar Bear walking, Sable Antelope, Sea Lion, Gorilla with pole, Wolf, Ostrich. (9 pieces)

Shown only as a catalog page:

#18Z to #23Z. Made by Hugar.
#18Z. Mammal House.
#20Z. Large Pool Enclosure.
#21Z. Monkey Hill.
#22Z. Animal House with Enclosure.
#23Z. Rock Pool.

Post-War Sets not illustrated

#26Z. Comprised of Young Hippopotamus, Young Rhinoceros, Young Indian Elephant, Young Giraffe, Lioness, Zebra, Panther, Gorilla with moveable arm, Eland Bull. (9 pieces)

#27Z. Comprised of Indian Elephant, Hippopotamus, Zebra, Crocodile, Panther, Baby Elephant, Chimpanzee, Gazelle, Young Giraffe, Sable Antelope, Gorilla with moveable arm, Guenon Monkey, Ostrich, Lion, Lioness, Young Rhinoceros, Bush Buck. (17 pieces)

Pre-War Only set numbers: 2Z, 4Z, 5Z, 6Z, 15Z, 17Z, 18Z, 19Z, 20Z, 21Z, 22Z, 23Z.

Pre- and post-War set numbers: 1Z, 3Z, 11Z.

Post-War Only set numbers: 24Z, 25Z, 26Z, 27Z.

Zoo Itemized Date of First Issue

Catalog Number	Issue Date
901-918	1930
919-933	1931
934-953	1932
954-967	1933
968	1934
969-971	1939
976	1947
978	1949
986-992	1952
993-994	1953

Catalog Number	Issue Date
1Z-4Z	1931
5Z, 6Z	1932
11Z	1934
15Z	?
17Z	1937
18Z-23Z	1940

Zoo Picture Packs

9001-9015	1949

Zoo tinplate advertising sign. $150-$200.

Don't say "Zoo," say "Britains' Zoo"

THERE are few things more interesting either to young or old, than a Zoo ; the variety of animals, each characteristically different from the others, majestic, ferocious or docile as the case may be, providing some thing new to hold the attention, and the whole surrounded by that indefinable attraction of the wild, appealing perhaps to some remote sense when the human being was more the hunted than the hunter.

We are, however, wandering from the point, as this is not intended to be a treatise on Natural History, but a short introduction to **BRITAINS LIMITED** Zoological Series, which is, as the name implies, a series of models taken wherever possible from life and reproduced with care as to detail, and which will give the "collector" the opportunity of having his or her Zoo at home and will, we hope, provide many long hours of pleasure and enjoyment, with perhaps some measure of education.

Zoo catalog introduction.

Pre-World War Two illustrated Zoo set label.

Zoo catalog outer cover. A foldout sheet in the back of this catalog giving a numerical listing of the Zoo Range also shows numbers #968 to #986. #968 was later allocated to Water Buffalo, #969 to #971 to Pandas. #972 to #974 were wholesale trade assortments, leaving #975 to #986 without an allocated item. This gave rise to speculation the other Zoo Animals were to be added. See, for example, the Roan Antelope illustration. #976 was later allocated to Sailor Boy to ride Camel, #978 to Baby Chimpanzee, and #986 to Panther, both of which were issued post-War.

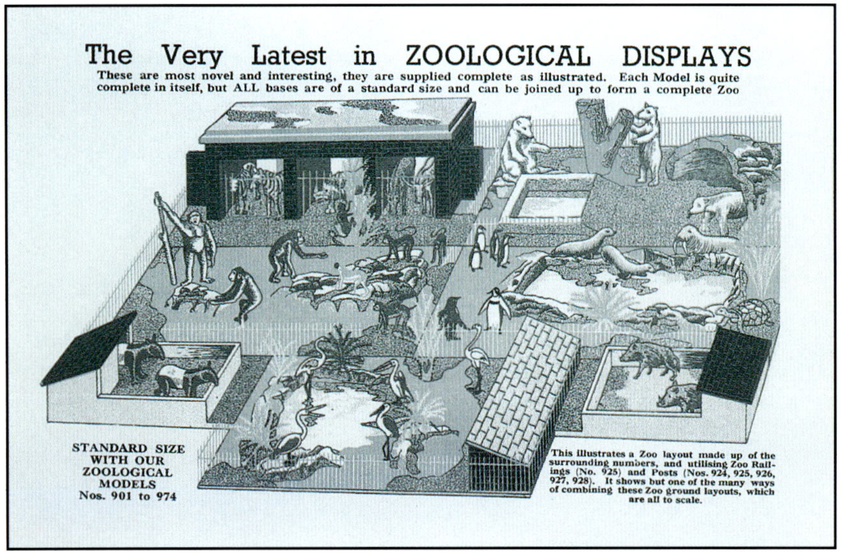

The Very Latest in ZOOLOGICAL DISPLAYS

These are most novel and interesting, they are supplied complete as illustrated. Each Model is quite complete in itself, but ALL bases are of a standard size and can be joined up to form a complete Zoo

STANDARD SIZE WITH OUR ZOOLOGICAL MODELS Nos. 901 to 974

This illustrates a Zoo layout made up of the surrounding numbers, and utilising Zoo Railings (No. 925) and Posts (Nos. 924, 925, 926, 927, 928). It shows but one of the many ways of combining these Zoo ground layouts, which are all to scale.

Catalog page showing Zoo display.

#902 Kangaroo. $25-$35 each.

#903 Penguin. $20-$30 each.

#901 Elephant. Rubber trunk. Pre-War only issue. $120-$150.

#901 Elephant. Lead trunk. Post-War box. $120-$150.

#904. Gibbon. Climbing Monkey. $25-$35 each.

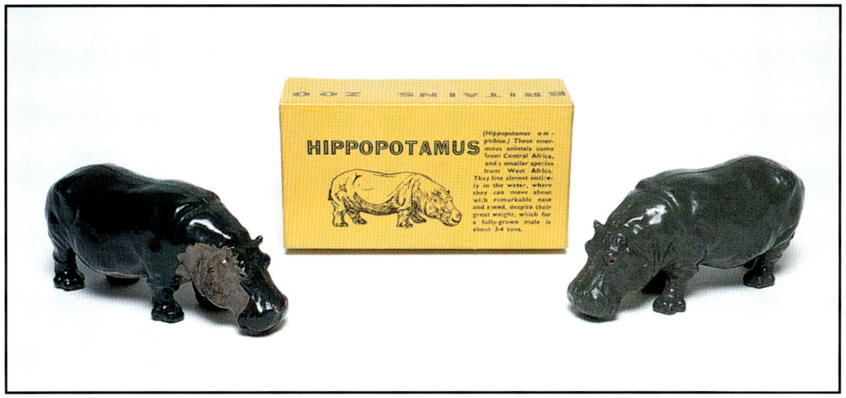

#905 Hippopotamus. Post-War box. $90-$100.

#908 Indian Rhinoceros. Post-War box. $90-$100.

#906 Gorilla. Moveable arm. $30-$35 each.

#908 Indian Rhinoceros and #905 Hippopotamus with examples of pre-War boxes. $110-$120 each.

#907 Zebra. $20-$25 each.

#909 Pelican. Beak open. $20-$25 each.

#910 Lion. $20-$30 each.

#911 Lioness. $20-$30 each.

#913 Pelican. Beak closed, wings spread. $20-$25 each.

Box art for adult Giraffe.

Bottom center: #912 Adult Giraffe. Pre-War issue only. $250-$280. Boxed example $350-$400.

Bottom right: #914 Polar Bear. Sitting up. $35-$40.

#915 Chimpanzee. $25-$35 each.

#917 Nile Crocodile. Post-War box. $90-$100.

#916 King Penguin. Front view. $20-$30 each.

#918 Bactrian Camel. $45-$55.

#916 King Penguin. Rear view. $20-$30.

#919 Coconut Palm. #920 Date Palm. $55-$65 each.

#921 Guenon Monkey. $20-$25 each.

#924 Gate with posts. $55-$65.

#922 Ostrich. $20-$25 each.

#925 Railings. Straight section. Issued with rounded tops instead of spikes post-War #994. $25-$30.

#923 Llama. $20-$25 each.

#926 Railings. Curved sections. Round top version (#994) post-War issue. $25-$30.

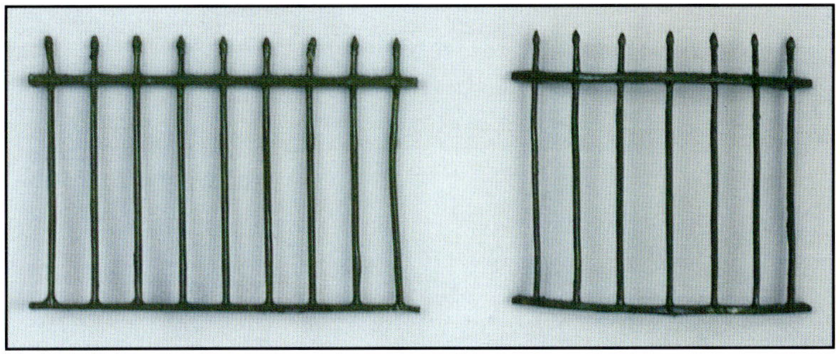

#925 Railing. Straight section spiked top. #926 Railing. Curved section spiked top. $25-$30 each.

#931 Zoo Keepers (see previous illustration.) Right: #932 Zoo Keeper, seated astride. Pre-War issue only against this catalog number. Issued only in boxed sets #24Z and #25Z post-War. Seated Keeper $80-$85.

#927 Standard post, two way straight. #928 Standard post, two way right angle. #930 Standard post, four way. #929 Standard post, three way (not illustrated). $30-$35 each.

#933 Eland Bull. $35-$40.

#931 Zoo Keepers. Tall walking and shorter standing version. Both allocated the same catalog number. Master Mold #719 allocated to tall version and #931 to shorter version. $55-$65 each.

#934 Brown Bear. Catalog description Brown Bear, but also found painted black. $25-$30.

#935 American Bison. $55-$60.

#936 Bear Cubs. Brown and black versions were produced. Master Mold #709 allocated to walking bear cub and #936 allocated to sitting bear cub. $25-$35 each.

#937 Giant Tortoise. $30-$35.

#938 Howdah for Elephant. #939 Boy or Girl for Howdah. Issued pre-War only against these catalog numbers. Post-War all items were issued only in boxed sets #24Z and #25Z. Master Mold #712 allocated to Howdah seat and #938 to Howdah saddle, #710 allocated to girl and #939 to boy. Howdah, $180-$200. Children, $75-$80 each.

Color variations of #939 boy and girl for Howdah from the Britains Archive. *Courtesy Christie's Auction House, London/Britains Archive Auction 1994.* Unique.

#941 Tiger Walking. $25-$35.

#940 Baby Hippopotamus. $25-$35.

#942 Wild Boar. $20-$30.

#943 Baby Camel. $20-$30.

#946 Stork. $20-$25.

#944 Baby Elephant. $25-$30.

#947 Flamingo. $20-$25.

#945 Sable Antelope. The figure on the right is a rare first version with shorter horns. Left: $30-$35; right: $40-$50.

#948 Wart Hog. $25-$30.

#949 Malay Tapir. $30-$35.

#952 Young Indian Elephant. $35-$45.

#950 Baby Kangaroo. $20-$25.

#953 Bactrian Camel with boy rider in sailor suit issued pre-War and post-War only in 1946. The boy was also issued post-War with individual catalog #976 and in Zoo display set #24Z. $90-$120.

#951 Baby Rhinoceros. $25-$30.

#954 Gorilla with pole. $30-$35 each.

#955 Wolf. $30-$35 each.

#956 Walrus. $25-$35 each.

#957 Red Deer. $30-$35 each.

#958 Young Crocodile. $35-$45 each.

#959 Young Hippopotamus. $30-$35 each.

#960 Young Rhinoceros. $30-$35 each.

#964 Sea Lion. $20-$30 each.

#961 Young Giraffe. $30-$35 each.

#962 Lion Cub sitting. $25-$30 each.

#965 Himalayan Bear sitting. $30-$35 each.

#963 Gazelle. $25-$30 each.

#966 Polar Bear walking and #967 Polar Bear standing. $30-$35 each.

#968 Water Buffalo. $30-$35.

#969 Giant Panda and two Panda cubs issued with the same #970 catalog number. #971 was a boxed set comprising all three and was issued pre-War only. The combination was revived post-War as a Picture Pack box #9011. Master Mold numbers for Panda cubs same as #936 Bear cubs. Panda, $30-$35. Cubs, $25-$30 each.

#976. Boy to ride camel in sailor suit. Issued post-War only as a single item and in #24Z display set. Pre-War available only as #953 with Bactrian Camel. $85-$95.

#978 Baby Chimpanzee. Although issued post-War for the first time with an individual catalog number, this item was originally issued in 1935 as "Jubilee" for the uncataloged BoBo and Jubilee set made for London Zoo. Master Mold #953 gives a pre-War identity number to this item. Because of the small size, the mold was converted from hollow cast to solid.; $45-$50.

#986 Panther (although looking more like a leopard), #987 Baboon, #988 Springbuck, #989 Bushbuck. All post-War issues only. $60-$65 each.

#990 Vulture. $65-$75.

#990 Vulture, #991 Cobra, the first showing Britains Archive, Christie's Auction lot tag, and #992 seated Tiger. The Tiger was first issued in 1951 as a circus Tiger #449B and later as a zoo figure in 1952. Vulture and Cobra, $65-$75 each. Tiger, $25-$30.

Set #1313. Eastern People included in this section because of the Noah's Ark figure connection. Noah and his wife having been used from this set. Post-War issue. Set. #1314 was a larger set pre-War only. *Courtesy Christie's Auction House, London/The Arnold Rolak Reference Collection.* $1,200-$1,500.

Eastern People. Left to right: Eastern Man with grey beard, Master Model #244B; Eastern Woman #245B, also used in set #1550 as Noah's wife; Eastern Boy #246B; Eastern Man #247B, also used in set #1550 as Noah; Eastern Shepherd #248B. This is the same casting as #244B with a brown beard. *Courtesy Philip Dean Reference Collection*. $80-$85 each.

#1550 Noah's Ark. Pre-War issue only. The top of the Ark is removable and the lower part serves as a box for the sets contents. *Photo courtesy Christie's Auction House, London/The Arnold Rolak Collection*. $6,000-$7,000.

#1550 Noah's Ark with animals. Pre-War issue only. *Photo courtesy Christie's Auction House, London/The Arnold Rolak Collection*. $6,400-$7,500.

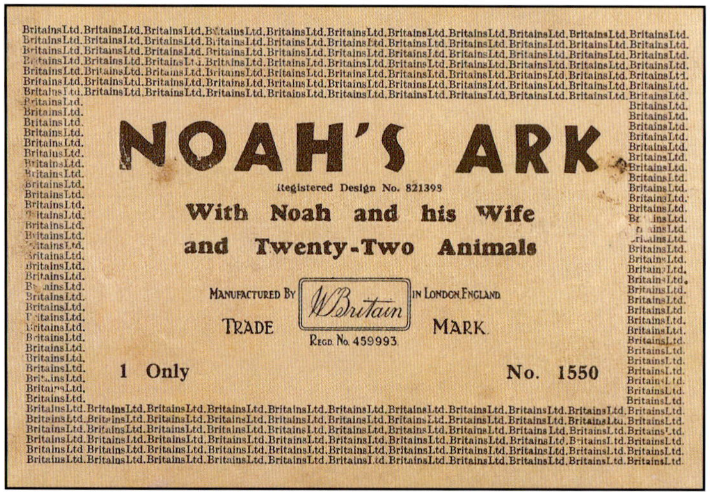

#1550 Noah's Ark box., lid and artwork. Pre-War issue only. *Photo courtesy Christie's Auction House, London/The Arnold Rolak Reference Collection.*

#1Z. Zoo set comprising Lion, Lioness, Zebra, Gorilla with moveable arm, climbing Monkey, Guenon Monkey, Pelican (beak open), Pelican (beak closed), and two King Penguins. Post-War box. *Photo courtesy Bob Bailey.* $200-$300.

#4Z. Zoo set box art. Pre-War only. *Courtesy Philip Dean Reference Collection.* $1,200-$1,500 complete set.

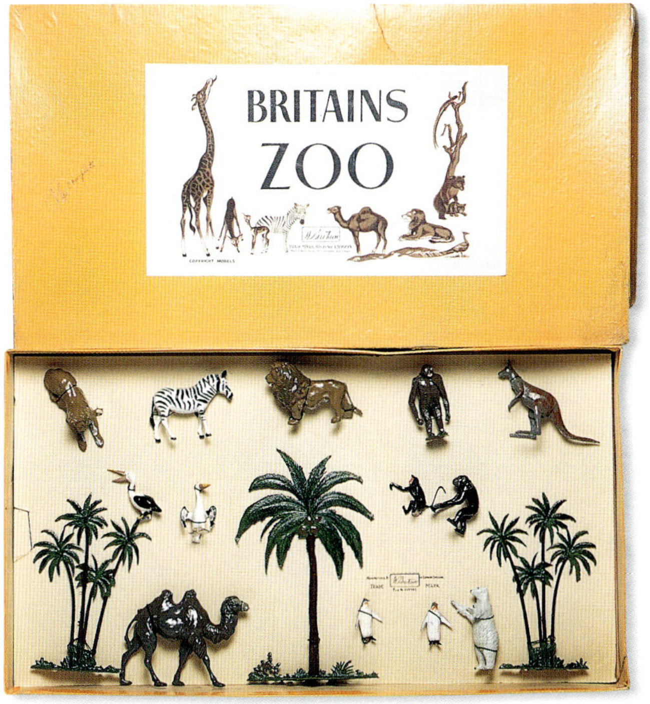

#3Z. Zoo set comprising Camel, Polar Bear seated, Lion, Lioness, Zebra, Kangaroo, Chimpanzee, Gorilla with moveable arm, Gibbon Climbing Monkey, Pelican (beak open), Pelican (beak closed), two Date palms and a Coconut palm. Post-War box. *Photo courtesy Bob Bailey.* $300-$400.

#4Z. Zoo set lower box. Rubber trunk Elephant, Indian Rhinoceros, Hippopotamus, Giraffe, Crocodile, Camel, Lion, Lioness, Coconut Palm and Date Palm. Pre-War only. *Courtesy Philip Dean Reference Collection.*

#4Z. Zoo set top tray. Pelicans, Penguins, King Penguins, Zebra, Llama, Ostrich, Polar Bear sitting, Gorilla, Gorilla with pole, Kangaroo and Gibbon climbing Monkey. Pre-War only. *Courtesy Philip Dean Reference Collection.*

#15Z. Zoo Set. This set never appeared in any Britains catalog. Comprising Llama, Himalayan Bear, Young Giraffe, Baby Camel, Baby Elephant, Brown Bear, and Sable Antelope. *Courtesy Ray Haradin.* $1,500-$1,700.

#11Z. Zoo set comprising Eland Bull, Sable Antelope, Guenon Monkey, Zebra, Walrus, Llama, Ostrich, Young Rhinoceros, Young Hippopotamus, Young Giraffe, Young Indian Elephant. Post-War box. *Photo courtesy Bob Bailey.* $400-$600.

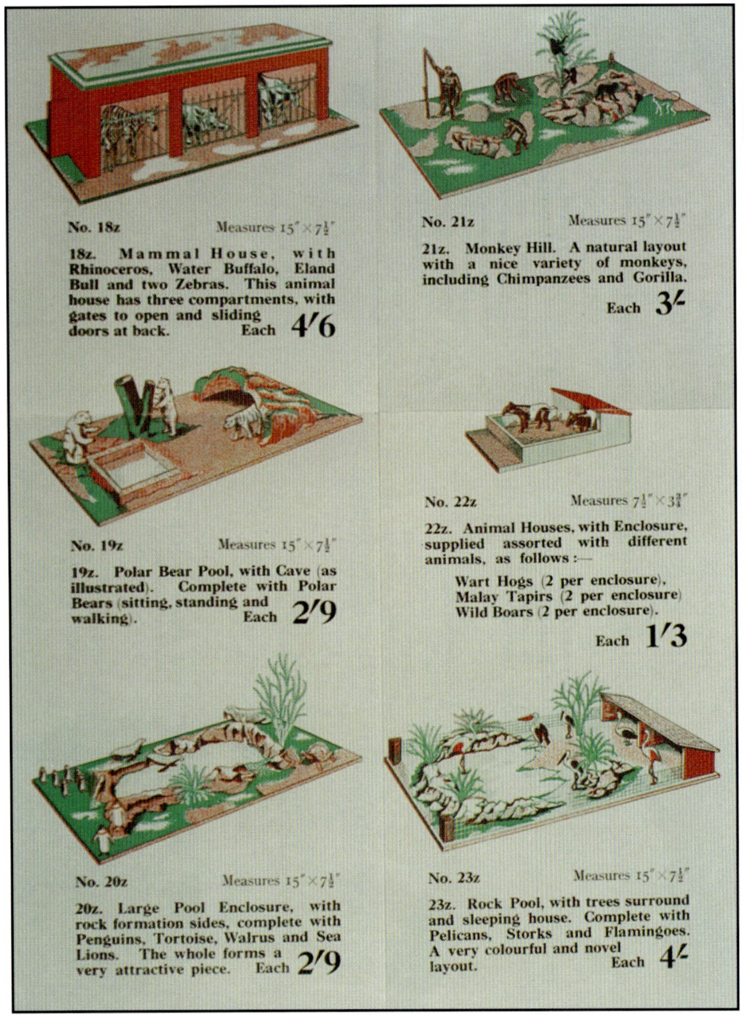

No. 18z Measures 15″ × 7½″

18z. Mammal House, with Rhinoceros, Water Buffalo, Eland Bull and two Zebras. This animal house has three compartments, with gates to open and sliding doors at back. Each 4/6

No. 21z Measures 15″ × 7½″

21z. Monkey Hill. A natural layout with a nice variety of monkeys, including Chimpanzees and Gorilla. Each 3/-

No. 19z Measures 15″ × 7½″

19z. Polar Bear Pool, with Cave (as illustrated). Complete with Polar Bears (sitting, standing and walking). Each 2/9

No. 22z Measures 7½″ × 3¾″

22z. Animal Houses, with Enclosure, supplied assorted with different animals, as follows:—

Wart Hogs (2 per enclosure).
Malay Tapirs (2 per enclosure).
Wild Boars (2 per enclosure).

Each 1/3

No. 20z Measures 15″ × 7½″

20z. Large Pool Enclosure, with rock formation sides, complete with Penguins, Tortoise, Walrus and Sea Lions. The whole forms a very attractive piece. Each 2/9

No. 23z Measures 15″ × 7½″

23z. Rock Pool, with trees surround and sleeping house. Complete with Pelicans, Storks and Flamingoes. A very colourful and novel layout. Each 4/-

Illustration from catalog dated 1940 showing enclosures: #18Z Mammal House; #19Z Polar Bear Pool; #20Z Large Pool Enclosure; #21Z Monkey Hill; #22Z Animal House and Enclosure; #23Z Rock Pool. Prices range from $2,000-$3,500 each depending on size of item.

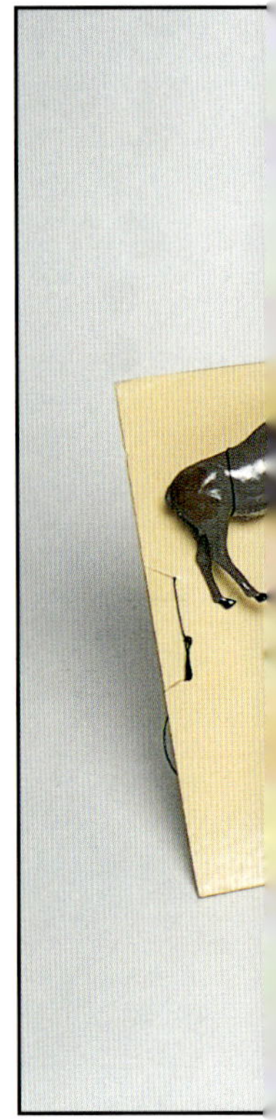

#19Z. Polar Bear pool and cave made by Hugar for Britains. Pre-War only. *Photo courtesy Christie's Auction House, London/The Arnold Rolak Reference Collection.* $2,000-$2,200.

#24Z. Zoo set inner tray with items still attached. *Photo courtesy Ray Haradin.*

#25Z. Elephant ride in original box. Post-War only issue. *Photo courtesy Bob Bailey.*

#24Z. Zoo set comprising Adult Elephant with keeper astride Howdah with boy and girl riders. Camel with boy in sailor suit, Baby Camel, Llama, Zebra, Eland Bull, walking Tiger, Warthog, Wild Boar, Giant Tortoise, Panda, two Panda cubs, and Date Palm. Post-War box. *Photo courtesy Bob Bailey.* $3,000-$3,500 complete set.

BoBo and Jubilee. Issued in 1935 to celebrate the birth of Baby Chimpanzee Jubilee in the British Monarchs Jubilee year. Using #915 Chimpanzee and uncataloged Baby Chimpanzee. Master Mold records allocate #953 to Baby Chimpanzee. Later reissued as a single item #978 post-War. $250-$350.

Bushbuck. Example of standard item and a figure bearing a "tag" with Made in England. The "tag" system was used pre-War for a short period during the 1930s. It is therefore not known how this post-War only issue was cast with a "tag". One theory is perhaps that the Bushbuck was due to be released pre-War and allocated one of the vacant pre-War zoo catalog numbers (see Roan Antelope), but was held back for post-War production. $90-$100 with tag.

Roan Antelope or Wildebeest. Uncataloged prototype. Possibly destined to be allocated one of the vacant numbers listed in the pre-War Zoo catalog chart. *Norman Joplin Collection*. Unique.

Zoo Picture Packs. Prices $65-$85 depending on components in each set.

Picture pack. Zoo box art work.

#9001. #944 Young Indian Elephant and #952 Baby Elephant.

#9002. #942 Wild Boar, #951 Baby Rhinoceros, and #960 Young Rhinoceros.

Top right: #9003. #940 Baby Hippopotamus and #959 Young Hippopotamus.

Center right: #9004. #902 Kangaroo and #950 Two Baby Kangaroos.

Bottom right: #9005. #934 Brown Bear and #950 Brown Bear Cubs sitting and walking.

ZOO

BRITAINS
ZOO

BRITAINS LTD., LONDON ENGLAND

BRITAINS LTD., LONDON ENGLAND

Contents No. 9011
1 GIANT PANDA,

Opposite page;
Top left: #9006. #943 Baby Camel and #949 Malay Tapir.

Center left: #9007. #903 Penguin, #966 Polar Bear walking, and #967 Polar Bear standing.

Bottom left: #9008. #907 Zebra and #963 Gazelle.

Top right: #9009. #933 Eland Bull and #946 Stork.

Center right: #9010. #910 Lion, #911 Lioness, and #962 Lion Cub.

Bottom right: #9011. #969 Giant Panda and #970 Baby Pandas walking and sitting.

#9012. #906 Gorilla, #915 Chimpanzee, #921 Guenon Monkey, and #978 Baby Chimpanzee.

#9013. #909 Pelican beak open, #913, Pelican beak closed, and #947 Two Flamingos.

#9014. #903 Penguin, #916 King Penguin, #956 Walrus, and #964 Sea Lion.

#9015. #941 Tiger walking and #992 Tiger sitting. The box lists this set as Tiger and Tigress although no indication is given as to which of the two items is the Tigress. Britains catalogs do not give a listing for Tigress.

20. Garden, 1930

Miniature Garden first appeared in 1930 and was produced up until 1941. Single items, along with set piece Garden display sets, were available, and a catalog devoted to Garden only was printed.

For the retail trade sale enhancing garden plans showing the various ways in which a garden could be designed, this was an unusually bold marketing plan previously not used by Britains in an attempt to promote the sale of this series.

- #1 MG to #27 MG were Miniature Gardening boxed sets.
- #30 MG was a large display set of Miniature Gardening and #31 MG was a fully modeled, three dimensional tree. See Catalog #58F in the Farm Range for an example of this item.

Miniature Garden

Games and Toys.

Britains' Wonder Line...
...is here !

"MINIATURE GARDENING"

Much has been said and heard about Britains' new series "Miniature Gardening." Britains Ltd. are renowned for producing winners, and in the new series they have excelled all expectations. "Miniature Gardening" consists of moulded flowers, plants, shrubs, etc., earth garden beds with grass borders ; crazy paving paths, lawns, rustic arches, walls, trellis fences, all in natural colours. The plants fix in holes in the ground, paths, lawns, beds and walls fit together in an endless variety of realistic designs. You cannot go wrong by stocking this line, with its infinite appeal and low price. Remember the Farmyard and Zoological Series ? Here is a line to beat even those. Pay us a visit. We will only be too glad to assist you in choosing your stocks and making your displays.

Britains' New City Showrooms

From May 1st we shall be occupying our new and spacious showrooms at Tredegar House, Milton Street, London, E.C. (2 minutes from Aldersgate Street Station) where you are cordially invited to call and inspect our range.

BRITAINS LTD., Lambton Rd., London, N.19 Telephone : ARCHWAY 3038

Please mention "GAMES AND TOYS" when writing to advertisers.

Miniature Garden Advertisement, 1931. *Courtesy "Games and Toys".*

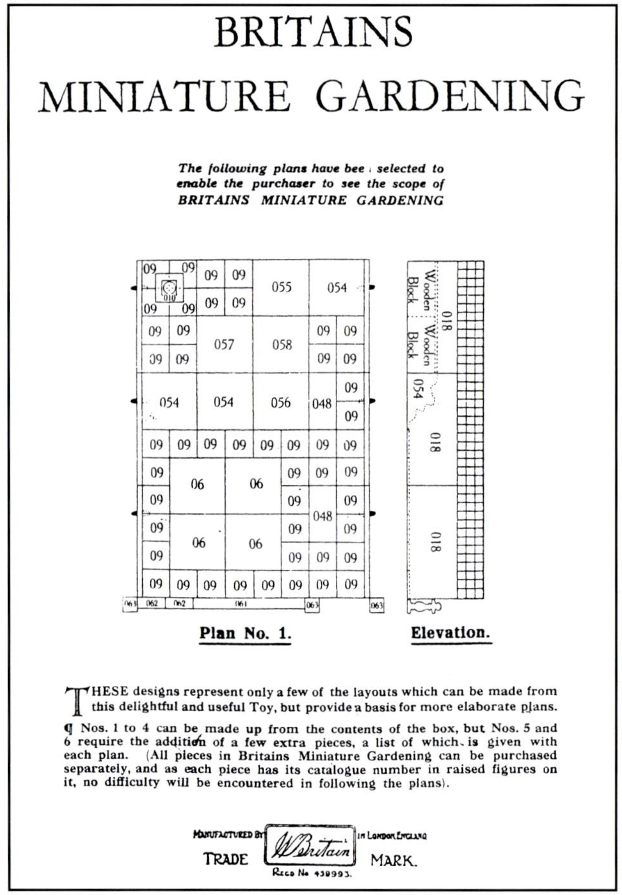

BRITAINS

MINIATURE GARDENING

The following plans have bee selected to enable the purchaser to see the scope of BRITAINS MINIATURE GARDENING

Plan No. 1. **Elevation.**

THESE designs represent only a few of the layouts which can be made from this delightful and useful Toy, but provide a basis for more elaborate plans.

¶ Nos. 1 to 4 can be made up from the contents of the box, but Nos. 5 and 6 require the addition of a few extra pieces, a list of which is given with each plan. (All pieces in Britains Miniature Gardening can be purchased separately, and as each piece has its catalogue number in raised figures on it, no difficulty will be encountered in following the plans).

MANUFACTURED BY W. Britain IN LONDON ENGLAND
TRADE MARK.
REGD No 439993.

Miniature Garden Plan 1.

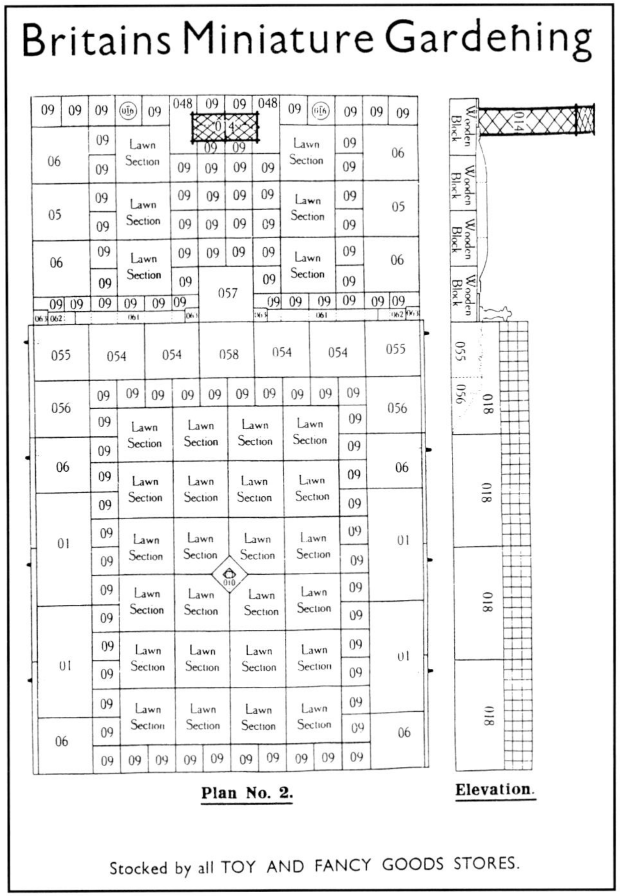

Miniature Garden Plan 2.

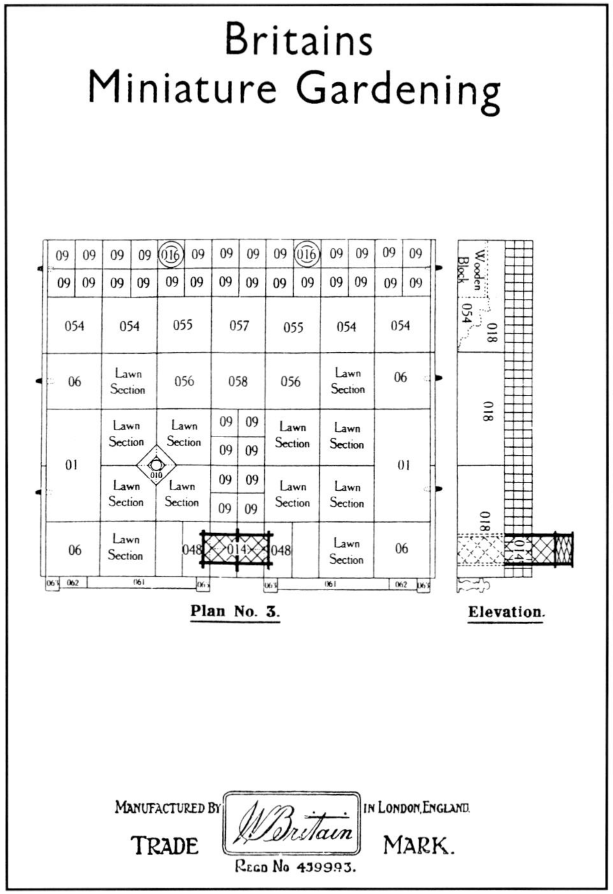

Miniature Garden Plan 3.

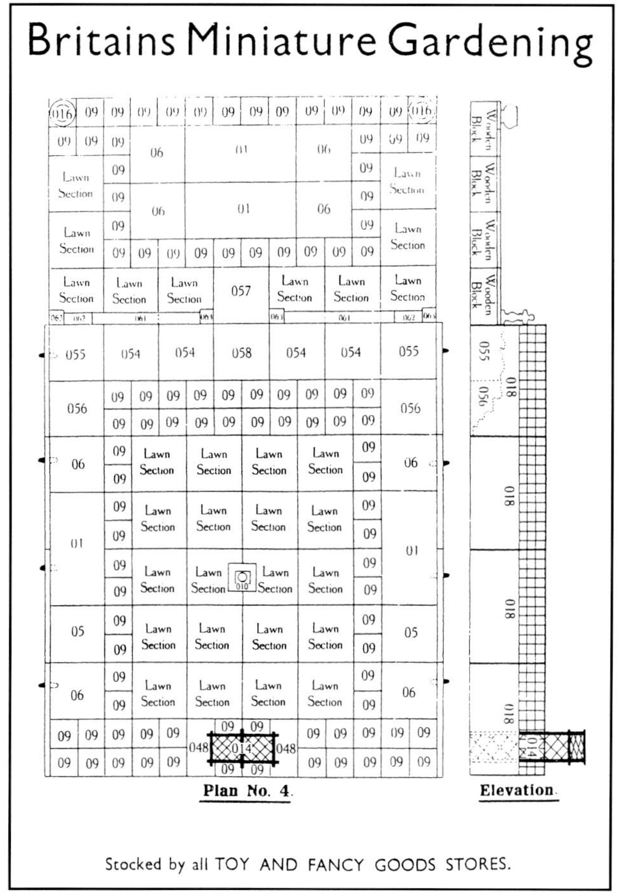

Miniature Garden Plan 4.

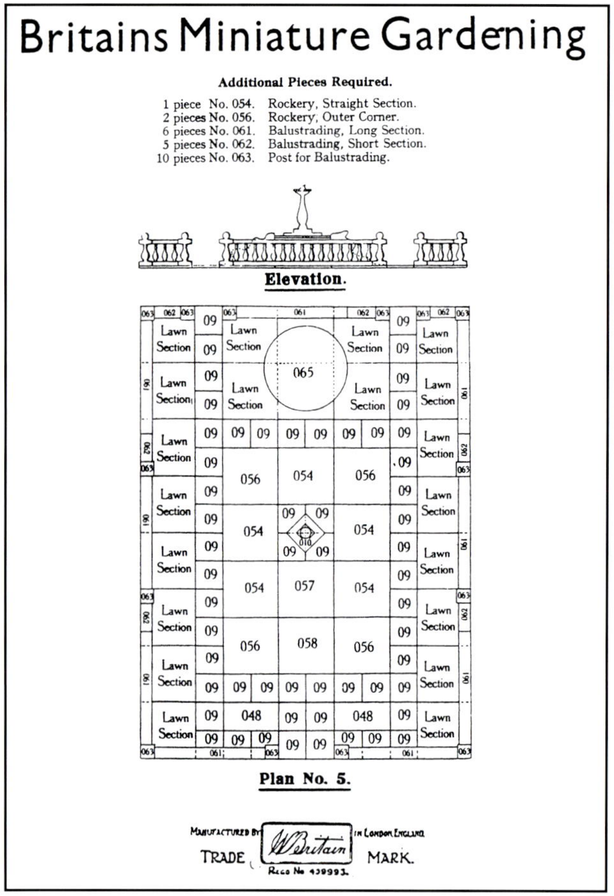

Britains Miniature Gardening

Additional Pieces Required.

1 piece No. 054. Rockery, Straight Section.
2 pieces No. 056. Rockery, Outer Corner.
6 pieces No. 061. Balustrading, Long Section.
5 pieces No. 062. Balustrading, Short Section.
10 pieces No. 063. Post for Balustrading.

Elevation.

Plan No. 5.

Manufactured by W. Britain in London England
TRADE MARK.
Regd No 439993.

Miniature Garden Plan 5.

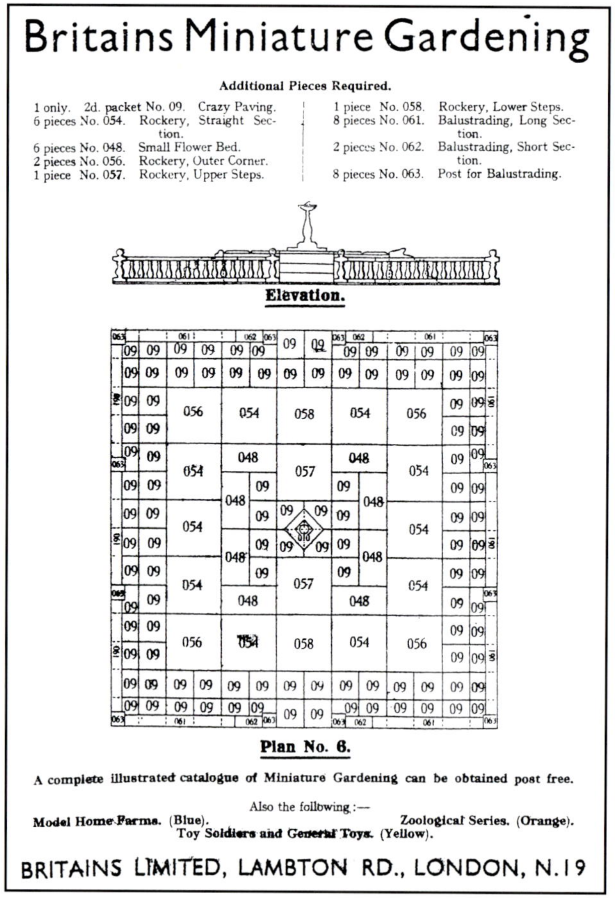

Britains Miniature Gardening

Additional Pieces Required.

1 only. 2d. packet No. 09. Crazy Paving.
6 pieces No. 054. Rockery, Straight Section.
6 pieces No. 048. Small Flower Bed.
2 pieces No. 056. Rockery, Outer Corner.
1 piece No. 057. Rockery, Upper Steps.
1 piece No. 058. Rockery, Lower Steps.
8 pieces No. 061. Balustrading, Long Section.
2 pieces No. 062. Balustrading, Short Section.
8 pieces No. 063. Post for Balustrading.

Elevation.

Plan No. 6.

A complete illustrated catalogue of Miniature Gardening can be obtained post free.

Also the following:—
Model Home Farms. (Blue). Zoological Series. (Orange).
Toy Soldiers and General Toys. (Yellow).

BRITAINS LIMITED, LAMBTON RD., LONDON, N.19

Miniature Garden Plan 6.

Garden wheelbarrow, Catalog #011. $15-$20.

Bottom left: Flower Beds. Top row, left to right:: Straight section with grass border, Catalog #01;, Finishing circular section with grass border, Catalog #02; Return square section with grass border, Catalog #03. Bottom row left to right: Return circular section with grass border, Catalog #04; Half straight section with grass border, Catalog #05; Finishing corner section with grass border, Catalog #06. $35-$40 each.

Bottom right: Post for stone wall, Catalog #07; Garden Roller, Catalog #08; Crazy paving section, two colors beige or grey, Catalog #09; Sundial, two colors grey or brown, Catalog #010. Left to right: $12-$15, $8-$10, $30-$35.

Stone wall, two colors, beige or grey, Catalog #012; Pergola, Catalog #013; Rustic arch, Catalog #014. Wall, $20-$25. Pergola, $75-$80. Rustic Arch, $80-$90.

Garden seat, Catalog #017. Interlaced board fence with trellis, Catalog #018. Seat, $15-$20. Fence, $30-$35.

Garden Seat, color variants. White Seat, $45-$55.

Center left: Mound, Catalog #015, a single holed earth mound to hold one flower. Also issued as a stand in the Circus Range against Master Model #1212B. The garden mound was allocated Master Mold #790. Colored vase, Catalog #016, beige, brown, and dark red color variants. Mound, $10-$12. Vas, $25-$35.

Bottom left: Colored vase, Catalog #016; Beige, brown and two red color variants. $25-$35 each.

Rambler Rose, Catalog #019. $80-$90.

Flowers. Left to right: Lobelia, Catalog #020; Geranium, Catalog #021; Torch Lily, Red Hot Poker, Catalog #022; Conifer, Catalog #023; Sunflower, Catalog #024; Poppy, Catalog #025; Lupin, Catalog #026; Half Standard Rose, Catalog #027; Bush Rose, Catalog #028; Aster, Catalog #029. Left to right: $10-$12, $12-$15, $12-$15, $10-$12, $25-$35, $15-$20, $15;-$20, $30-$35, $20-$30, $10-$15.

Flowers. Left to right: Hollyhock, Catalog #030; Antirrhinum, Catalog #031; Dahlia Double, Catalog #032; Dahlia Single, Catalog #033; Gladioli, Catalog #034; Wallflower, Catalog #035; Foxglove, Catalog #036; Chrysanthemum, Catalog #037; Full Standard Rose, Catalog #038; Delphinium, Catalog #039. Left to right: $15-$20, $10-$15, $15-$18, $15-$18, $10-$12, $12-$15, $10-$12, $12-$15, $30-$35, $12-$15.

Flowers. Left to right: Hyacinth, Catalog #040; Tulip, Catalog #041; Crocus, Catalog #042, Snowdrop, Catalog #043; Daffodil, Catalog #044; Narcissus, Catalog #045, Sweet Alyssum, Catalog #046. Left to right: $10-$15, $12-$15, $10-$12, $10-$12, $12-$15, $12-$15, $10-$15.

Square tub, Catalog #047; Small flower bed, Catalog #048. This item could be used in the main as support for Pergola and Rustic Arch. Tub, $15-$20. Flower Bed, $10-$12.

Lawn section made of cork, Catalog #049; Lawnmower, Catalog #050, for use with Lawnmower Man, Catalog #051. Lawn, $8-$10 per section. Lawnmower, $65-$70.

Man for Lawnmower, Catalog #051, issued in the Farm Series pre-War only (1940-41) as Catalog #659. Man for Garden Roller (not illustrated). See Farm Chapter, Catalog #680. This is the same figure as the Lawnmower Man with adapted arms. The Rollerman's hands/arms when placed behind the body, hook onto the roller handle to pull (see Catalog #08), while the Lawnmower Man's hands/arms are outstretched to push the lawnmower. $85-$90.

Span roof greenhouse and box, Catalog #053. This item appeared briefly in the Farm Range pre-War as Catalog #112F. *Photo courtesy of Debbie Bednarek*. $350-$400.

Span roof greenhouse box art variant with printer's flower design. *Courtesy Christie's Auction House, London.*

Rockery. Straight section, Catalog #054, two variants. $30-$35 each.

Rockery (see following illustration for different casting versions). Straight section, Catalog #054; Inner return corner, Catalog #055; Outer corner, Catalog #056; Upper steps, Catalog #057; Lower Steps, Catalog #058. $30-$35 each.

Flower pots, three sizes, Catalog #059, with original packet. *Courtesy Philip Dean Reference Collection*. $65-$70.

Wooden blocks for terracing, Catalog #060. $15-$20 each.

Cold Frame, Catalog #064. $65-$75.

Balustrading. Long section, Catalog #061; Short section, Catalog #062; Balustrade Post, Catalog #063. Long section, $25-$30. Short section, $20-$30. Post, $25-$35.

Round Flower Bed without grass border, Catalog #065; Square Flower Bed without grass border, Catalog #066. $65-$75 each.

Lily Pond with Nymph, Catalog #067. Although a separate item, the Nymph which "plugs" into the center of the pond was not given a catalog number. The Master Mold #778 gives this item its own identity reference. $180-$250.

Hose on reel, Catalog #068; Seed Box, Catalog #069. Catalog #070 was allocated to Man for Garden Wheelbarrow (not illustrated). See pre-War Catalog #678 in the Farm Range. Hose Reel, $50-$55. Seed Box, $35-$40.

#2 MG Miniature Gardening Flower Bed and Flower Combinations with box. *Courtesy Philip Dean Reference Collection*. $300-$400.

Water Lily; no catalog number was allocated to this item. It is recorded in Britains Factory Master Mold List as #779. $60-$70.

#4 MG Miniature Gardening Set comprising stone walls, conifers, flowers and flower beds. *Courtesy Philip Dean Reference Collection.* $400-$500.

#13 MG Miniature Gardening Set box art.

#10 MG Miniature Gardening Set with box. *Courtesy Philip Dean Reference Collection.* $500-$600.

#28 MG Garden Shelter with box, listed briefly pre-War in the Farm Series as Catalog #113F. *Courtesy Christie's Auction House, London/The Arnold Rolak Reference Collection.* $700-$800.

#29 MG Miniature Gardening box art showing paper mache cottage (see Catalog #44F in Farm Section). This set also included various Flint Wall pieces from the Farm Range.

Garden display showing the use of lily pond, rockery, and terracing. *Courtesy Philip Dean Reference Collection.*

Garden display showing use of flowers, lawn, crazy paving, walls, and interlaced fence. *Courtesy Philip Dean Reference Collection.*

#35 MG catalog illustration from 1937 showing the largest Garden display.

Miniature Garden items in their original packets. Top row, left to right: Mound #015; Paving #09; Lawn #049. Second Row, left to right: Torch Lily #022; Conifer #023; Sunflower #024; Half Standard Rose #027. Third Row, left to right: Hollyhock #030; Dahlia, single #033; Gladioli #034; Foxglove #036. Bottom Row, left to right: Chrysanthemum #037; Hyacinth #040. $65-$85 per packet.

Components that made up Span Roof Greenhouse #053 and Garden Shelter #28 MG with boxes. *Photo courtesy Christie's Auction House London.*

21. Outside Manufacturers, 1932

Advertising, Promotional, Character, and Experimental Issues

Items made by Britains for various companies provided much employment during the 1930s, although this practice started in 1924 with the issue of the Sarsons Lamb Figure. Madame Tussaud's Walt Disney, Cadbury's and J. Dille's Buck Rogers, etc. items accounted for a large amount of the production carried out between 1924 and 1941, both in London and the two regional sub-factories in Colne, Lancashire, and Liverpool.

Nestle's Cow

One of the first items produced was the Map of the World Cow or "Worlds Cow" as it was called by Nestle's. This figure was retooled from the Britains farm cow and a number of variations have been found to exist. The cow was a premium or gift given away by Nestle's at the 1924 British Empire Exhibition held at Wembley in London. The casting contained the words "The Worlds Cow" on one side and "Nestle's Milk" on the other, the outline of the map of the World was engraved in the casting and the countries painted in black against the white background of the cow. A rarer version exists without the words printed on, as does a brown and white paint variant.

Neither the Britains Factory Records or the Smallwood notebooks show a Mold or Master Model # for this item. The Master Mold list does contain the description Nestle's Cow; however, the reference # has been obliterated.

Felix the Cat

Dennis Britain, in an interview given in 1986, confirms the existence of a Felix designed by him in the late 1920s at the height of the cartoon character's popularity. No record appears in the Britains Factory records for either a Master Mold or Master Model number. As the Smallwood notebooks date from 1932, then no record appears in this document either. The Felix commonly found and attributed to Britains bears no markings.

Sarson's Vinegar

	Master Mold #	Mold Made
Sarson's Lamb	208	22 June 1932

Royal National Lifeboat Institute

	Master Mold #	Mold Made
Lifeboatman 65mm	826	16 February 1934

The entry in the Smallwood note books indicates that the base for this figure was originally square. The item is found painted in both pre- and post-War colors and was certainly issued as a promotional item during the 1950s. No recorded sighting has been made of a square-based version.

Cadbury's Cococubs

Lead animals issued in tins of Cadbury's Cocoa, 1934. This scheme was by far the most successful of the promotional items produced by Britains and warrants a chapter of its own. (See next chapter.)

Madame Tussaud's

Madame Tussaud's is a famous Wax Museum in central London. The Britains figures were available from the museum gift shop during the 1930s.

	Master Mold #	Tussaud's #	Mold made
Cinderella seated with broom	509 (Cinderella) 1077 (broom)	T5	5 June 1934
Red Riding Hood	510	T3	18 June 1934
Henry VIII Madame	511	T1	11 May 1934
Tussaud Bust	512	T2	6 June 1934

Queen Elizabeth I	513	T4	28 June 1934
Charles Dickens Bust*	514		29 March 1934
Micawber bust wearing hat*	515 (bust) 1078 (hat)		11 April 1934 / 25 April 1934
Scrooge Bust*	516		19 March 1934

*These three were designed as bottle stopper

Prince of Wales in Army Uniform (Master Model #1132B)	521	T6	27 August 1934

This item was re-released as King George VI for the Royal visit to South Africa in 1947 and was commissioned by Cape Town department store Stuttaford & Company, Ltd. Using mold #521 with a newly designed base and was issued in a blue presentation box. (See Royalty chapter)

John Dille and Company, Ltd. USA

	Master Mold #	Master Model #	Mold Made
Buck Rogers	973	321B	10 January 1935
Killer Kane	974	323B	31 January 1935
Ardalla	975	324B	26 January 1935
Robot (Mekkano Man)	976/979(arms)	326B	6 March 1935 / 12 March 1935
Wilma	977	322B	8 February 1935
Dr. Huer	978	325B	12 March 1935

All figures were available as single items through Woolworth Stores in the USA and were given as premiums when sending a number of Cream of Wheat tokens into J. Dille.

Britains Factory records show two catalog set numbers for these figures. Set #1415 has 6 pieces and set #1416 has 10 pieces, including 4 additional Robots. There is no recorded discovery of an original box for either set.

Robot Prototype. The Britains Master Model list number is 328B. No Master Mold number is recorded.

Johnnie Walker

	Master Mold #	Mold Made
A miniature Talisman	1052	8 August 1935

One-and-a-half inches high attached to a wire and secured to the cork in whisky bottles made by the Johnnie Walker Whisky firm. Britains were to continue this connection after World War II by issuing large Walker plastic figures of the Johnnie Walker man in two versions.

Buchanan's Whisky

Scottie Dog sitting	1027 (prototype only)	7 Sept. 1936
Scottie Dog standing	1028	30 June 1936

Scottie Dog Talisman for Buchanan's Whisky bottles. All dogs were issued in both black and white.

Disney Characters

	Master Mold #	Catalog #	Mold Made
Mickey Mouse	1376/1377	16H	24 Dec. 1937
Minnie Mouse	1378/1379	17H	6 May 1938
Donald Duck	1380/1381	19H	13 Nov. 1937 / 14 Dec. 1937
Goofy	1382/1383	21H	31 Jan. 1938
Pluto	1384/1385	18H	4 Jan. 1938 / 7 Jan. 1938
Clarabelle Cow	1386/1387	20H	20 Jan. 1938 / 18 Jan. 1938

All the above had removable, moveable heads. The dates given are those when each mold was made, the first being the body mold, the second the head mold. Those with one date indicates both head and body molds were made on the same day. It is interesting to note that the Clarabelle Cow head mold was made before the body mold.

These items retailed as single items and also in boxed set #1645 and set #1644 Cinema set.

Walt Disney

	Master Mold #	Catalog #	Mold Made
Snow White	1019	548B	30 May 1938
Sleepy	1020	541B	12 May 1938
Happy	1021	542B	19 May 1938
Sneezy	1022	543B	29 July 1938
Bashful	1023	544B	3 May 1938
Grumpy	1024	545B	6 May 1938
Doc	1025	546B	27 May 1938
Dopey	1026	547B	1 June 1938

The dates given are those dates when the mold for each figure was first produced.

The above all came in a boxed set #1654 and also with two versions of wooden cottages produced by Hugar for Britains, a small cottage and larger cottage with baseboard and bridge with stream. Both were allocated #1659. The Snow White and Dwarfs were sometimes also available with the larger cottage boxed set.

Pitchford-Velos

Father Christmas pencil sharpener small	898	9 June 1937
Father Christmas pencil sharpener large	587	September 1937, no specific date in Small wood

Tower of the Empire

Mold made 4 May 1938. No Model # recorded. Souvenir for the Empire Exhibition held in Glasgow Scotland

Lambeth Walk Dancing Couple

London stage musical "Me and My Girl" featuring the song Lambeth Walk and starring Lupino Lane.

	Master Mold #	Master Model #	Mold Made
Dancing Boy	507	586B	1938
Dancing Girl	508	586B	1938

Figures were 45mm in height and were designed to link arms, issued 1939 only, Catalog number same as Master Model #586B.

Sir Kreemy Knut,

	Master Mold #	Mold Made
The "Sharps Toffee Man,"	545	?

Post-War Issues

Post-World War II the following items were made:

The National Association of Girls Clubs

	Master Mold #	Mold Made
Mr. Whatno	1131B	1946

Painters Set Catalog #1495

	Master Mold #
Painter to carry ladder (2)	1312
Painter to stand on ladder with paint brush in moveable arm	1313
Ladder	1314

Based on an advertisement for Ripolin Paints

Toy Town Soldiers/Grenadiers

	Master Mold #
Toy Town Soldier/Grenadier at attention	1298
Toy Town Soldier/Grenadier Marching	1299
Toy Town Soldier/Grenadier Mounted Officer	1300

Thought to be experimental figures for the purposes of testing the spray painting of Britains figures in the early nineteen fifties. The figures first came to light in public during a British Model Soldier Society Meeting in London in 1986, when a box set was shown at the meeting. Examples of the two foot figure were found in the Britains Archive and offered for sale at the Archive auction in 1994. The Toy Town Soldiers featured heavily in Christie's Auction House advertising of the Archive Auction. Several paint variations exist in Red, Yellow and Grey tunics. The items were distributed in North America for a limited trial period during the nineteen fifties.

World Cow showing "The World Cow" printed on side. $180-$220.

World Cow showing "Nestle's Milk" printed on reverse side. $180-$220.

World Cow. Two examples, with and without printing. Without printing, $200-$250.

Very rare World Cow. Brown and white paint version. *Courtesy Betty Wheeler.* $500-$600.

World Cow bulk pack. *Courtesy Christie's Auction House, London/ The Arnold Rolak Reference Collection.* $2,400-$2,600.

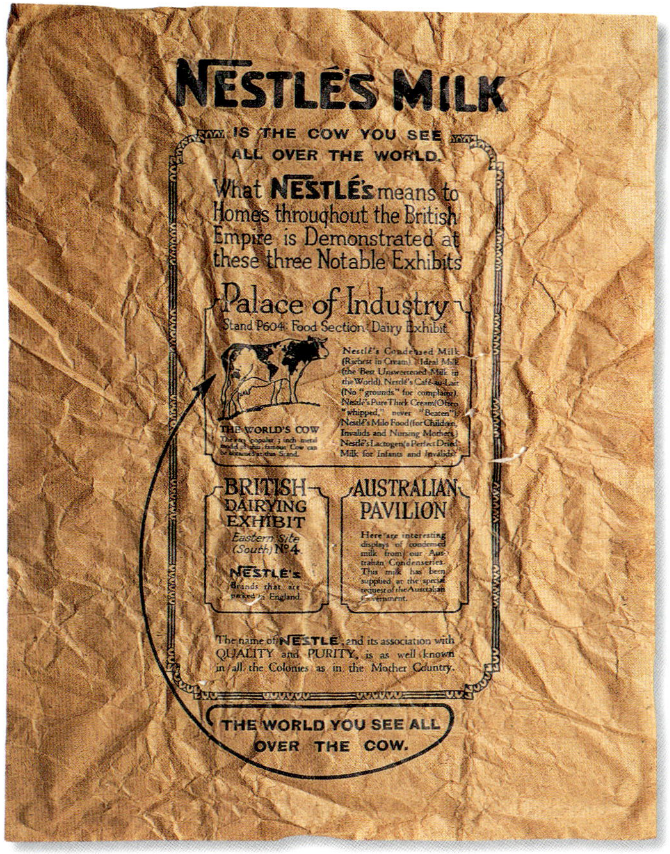

World Cow wrapper. 1924, Nestle's. British Empire Exhibition wrapping paper for World Cow. *Courtesy Dr. Rob Wilson.*

Cococub Group. Illustration shows one of each animal produced. See Cococub Chapter. *Courtesy Dudley Chignall.* Prices: See Cococub chapter.

Felix the Cat. *Courtesy Christie's Auction House, London/The Arnold Rolak Reference Collection.* $400-$500.

Sarson's Lamb. *Courtesy Christie's Auction House, London/The Arnold Rolak Reference Collection.* $40-$50.

Cococub Boys. See Cococub Chapter. *Courtesy Dudley Chignall.* Prices: See Cococub chapter.

Top left to right: Madame Tussaud's Henry VIII. *Courtesy Christie's Auction House, London/The Arnold Rolak Reference Collection.* $150-$180.

Madame Tussaud's Queen Elizabeth I. *Courtesy Christie's Auction House, London/The Arnold Rolak Reference Collection.* $150-$180.

Madame Tussaud's Red Riding Hood. *Courtesy Christie's Auction House, London/The Arnold Rolak Reference Collection.* $300-$350.

Madame Tussaud's Cinderella. *Courtesy Christie's Auction House, London/The Arnold Rolak Reference Collection.* $300-$350.

Left: Madame Tussaud's Madame Tussaud Bust. *Courtesy Christie's Auction House, London/The Arnold Rolak Reference Collection.* $150-$180.
Right: Madame Tussaud's Prince of Wales. *Courtesy Christie's Auction House, London/The Arnold Rolak Reference Collection.* $2,000-$2,500.

Above: Lifeboatmen. *Courtesy Christie's Auction House, London/The Arnold Rolak Reference Collection.* $70-$80 each.
Right: Charles Dickens Bust. *Courtesy Christie's Auction House, London/The Arnold Rolak Reference Collection.* $400-$500.

Johnny Walker Whisky talisman. $30-$40.

Buck Rogers. Manufactured under license from J. Dille. Left to right: Buck Rogers, Wilma, Killer Kane, Ardalla, Dr. Huer, and Mekkano-Man Robot. *Courtesy Christie's Auction House, London/The Arnold Rolak Reference Collection.* $350-$375 each. Mekkano Man Robot $400-$450.

Joseph Chamberlain Bust, souvenir from 1906 catalog. *Courtesy Christie's Auction House, London/The Arnold Rolak Reference Collection.* $250-$350.

Buck Rogers. Prototype robot, "Mechanica". *Courtesy Christie's Auction House, London/The Arnold Rolak Reference Collection.* Unique.

Pitchford-Velos Santa Claus pencil sharpeners. *Courtesy Christie's Auction House, London/The Arnold Rolak Reference Collection.* Large, $300-$350. Small, $180-$250.

Buchanan's Black and White Whisky Dogs. *Courtesy Christie's Auction House, London/The Arnold Rolak Reference Collection.* $65-$85 each.

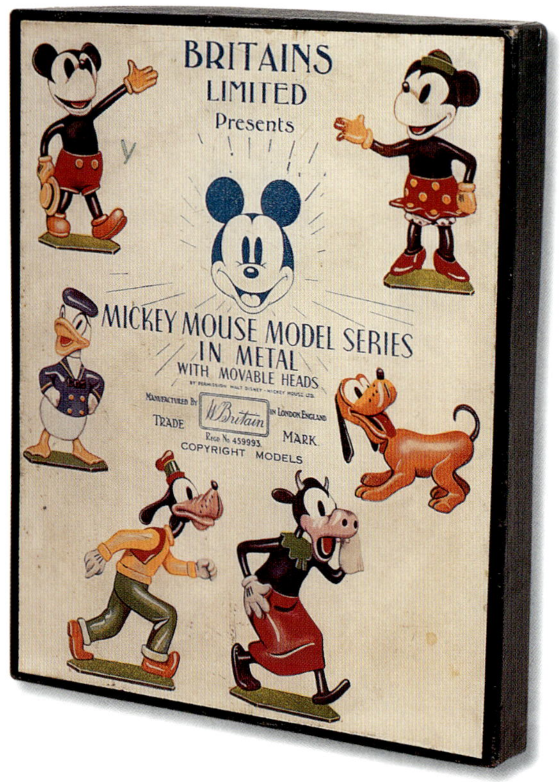

Mickey Mouse boxed set lid. Set #1645. *Courtesy Peter Kirk.*

Buchanan's Black and White Whisky Dog talisman. *Courtesy Christie's Auction House, London/The Arnold Rolak Reference Collection.* $50-$60.

Mickey Mouse set #1645 showing partitioned box. *Courtesy Peter Kirk*. $8,000-$10,000.

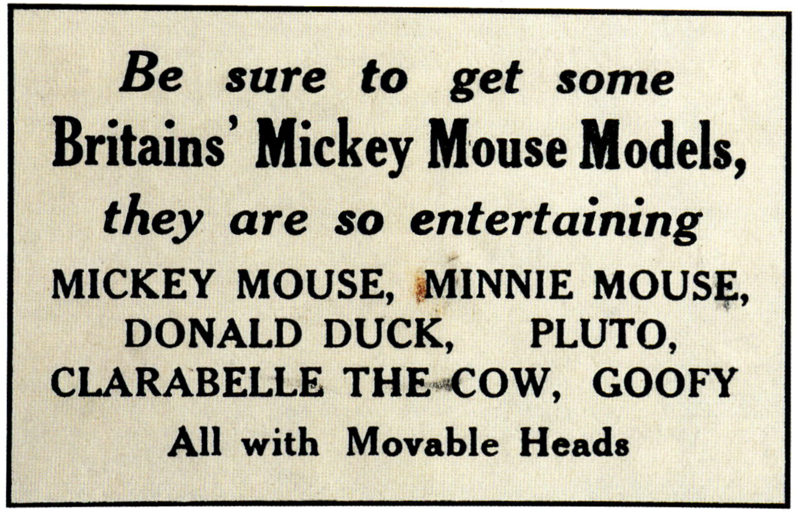

Advertising label which appeared inside Snow White box set for the Disney figures. *Courtesy Philip Dean.*

Mickey Mouse set #1644. Cinema box.

MANUFACTURED BY W Britain IN LONDON ENGLAND
TRADE MARK.
REGD No 439993.

Measures 12" × 8½"

No. 1644 MICKEY MOUSE AT THE CINEMA s. d.
This is an ingenious display box, containing: Mickey Mouse, Minnie Mouse, Pluto, Donald Duck, Clarabelle and Goofy, and so designed to represent a cinema show. The characters are arranged behind the title screen which can then be drawn up, so that the figures then appear as seen at the Cinema. Provides lots of fun ... 6 0

Snow White and Seven Dwarfs box set #1654. *Courtesy Philip Dean Reference Collection*. $2,000-$3,000.

Disney Characters, Left to right: Mickey Mouse, Minnie Mouse, Pluto, Donald Duck, Clarabelle Cow, and Goofy. *Courtesy Christie's Auction House, London/The Arnold Rolak Reference Collection*. $350-$400 each. Clarabelle Cow, $450-$500.

Snow White set. Left to right: Sleepy, Happy, Sneezy, Bashful, Grumpy, Doc, Dopey and Snow White. Dwarfs, $60-$80 each. Snow White, $100-$120.

Snow White. Large version cottage with stream and baseboard. The wooden cottage and the smaller version shown in the following illustration were manufactured by Hugar for Britains. *Courtesy Christie's Auction House, London/The Arnold Rolak Reference Collection.* With figures $8,000-$9,000.

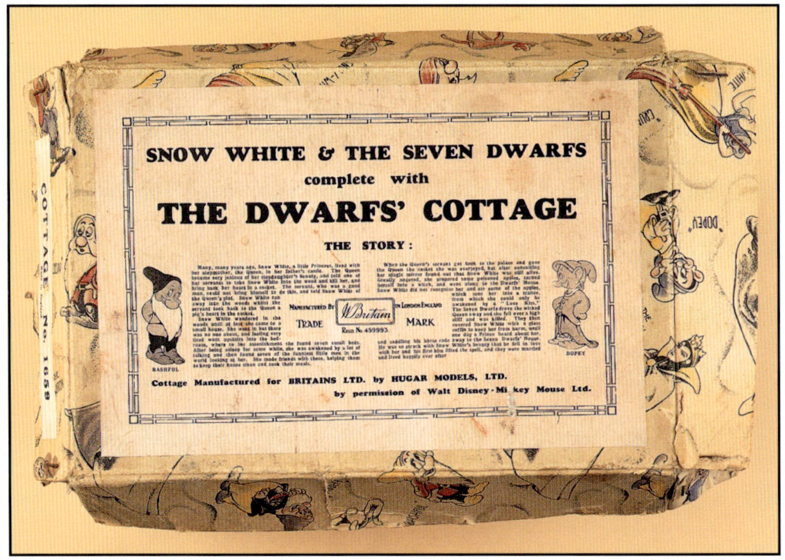

Snow White and the Seven Dwarfs complete with Dwarf's Cottage. Box lid. Catalog #1659.

Snow White cottage. Figures and box, Catalog #1659. *Courtesy Christie's Auction House, London.* $8,000-$9,000.

Prototype Felix the Cat. Wax master. This item was not put into production. *Courtesy Christie's Auction House, London/The Arnold Rolak Reference Collection.* Unique.

Prototype Mickey Mouse. Wax master. This item was not put into production. *Courtesy Christie's Auction House, London/The Arnold Rolak Reference Collection.* Unique.

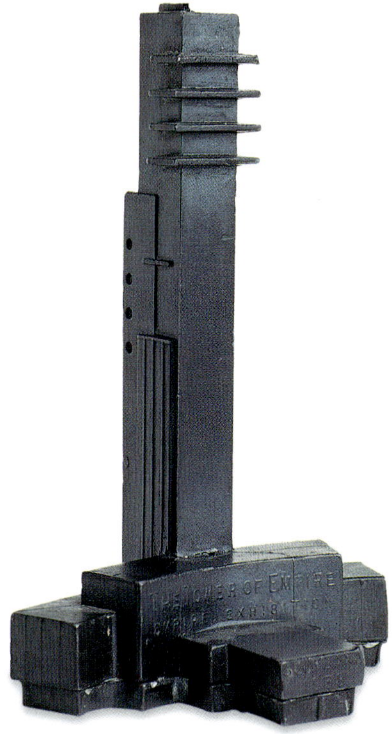

Tower of Empire. 1938 Glasgow Empire Exhibition. *Courtesy Christie's Auction House, London/The Arnold Rolak Reference Collection.* $250-$300.

Below: Prototype Pip, Squeak, and Wilfred. This item was not put into production. *Courtesy Christie's Auction House, London/The Arnold Rolak Reference Collection.* Unique.

Lambeth Walk. Dancing couple inspired by the London stage musical "Me and My Girl". *Courtesy Christie's Auction House, London/The Arnold Rolak Reference Collection.* $1,000-$1,500 each pair.

Lambeth Walk. Dancing couple with front and rear views. *Courtesy Philip Dean Reference Collection.* $1,000-$1,500 each pair.

Toy Town Grenadiers. *Courtesy Christie's Auction House, London/The Arnold Rolak Reference Collection.* $300-$400 each.

Toy Town Grenadiers. *Courtesy Christie's Auction House, London/Britains Archive Auction 1994.* $300-$400 each.

Sir Kreemy Knut, Sharp's Toffee Man with trade box 22XX. *Courtesy Christie's Auction House, London/The Arnold Rolak Reference Collection.* Unique.

Sir Kreemy Knut, Sharp's Toffee Man. $150-$180.

Toy Town Grenadiers box set. *Courtesy Old Toy Soldier Newsletter, Steve Sommers.* Unique.

Whitbread Tankards. *Courtesy Christie's Auction House, London/The Arnold Rolak Reference Collection*. $30-$40 each.

Marching Guardsman. The figure has a pin inserted into the hip which is attached to the handle in the base. When pushed back and forth; the Guardsman's leg moves. A prototype that was not put into production. *Courtesy Christie's Auction House, London/Britains Archive 1994*. Unique.

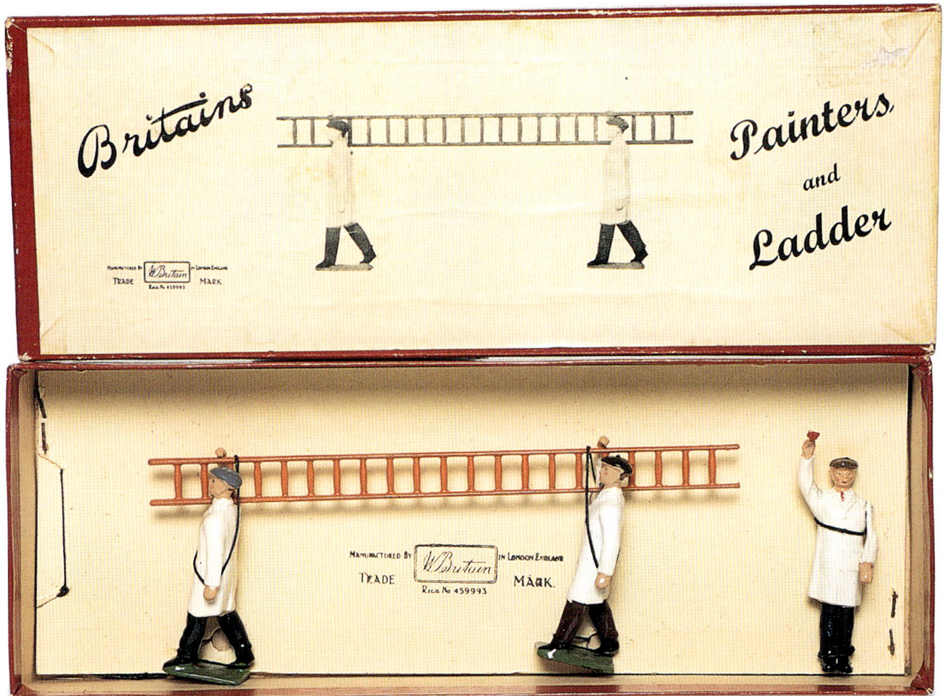

Ripolin Painters. The figures contained in this set Catalog #1495 were produced as advertising items for Ripolin Paints. The figures were then introduced into Britains Standard Range with individual catalog numbers allocated to the Farm Series. $400-$450.

Mr. Whatno. Produced post-War as a fund raiser for the Associations of Girls Clubs. *Courtesy Christie's Auction House, London/The Arnold Rolak Reference Collection*. $60-$80.

Marching guardsman advertising bookmark for Britains Archive auction, 1994. *Courtesy Christie's Auction House, London/Britains Archive 1994*.

22. Cococubs, 1934

In a promotional scheme that ranks not only as the most innovative and successful promotion involving children's toys during the Pre-World War II period, but also as a landmark in Britains lead figure production, the two firms entered into an agreement which resulted in the production of lead animals to be given away in tins of Cadbury's Bournville Cocoa. The paper labels on each tin carried the motto "A Toy in Every Tin." The joint scheme, which first came to the attention of the public in October, 1934, required the opening of two new factories in England to cope with the production of Cococubs, as they became known. These factories were located in Colne, Lancashire and Liverpool, Merseyside. Much Cococub production was carried out there, along with production of promotional items made in the 1930s for numerous other companies. *(see previous chapter)*

The following lists give details of the animals produced, which were made in two heights: 1.5" and 2.25". They were wrapped in greaseproof paper to separate the lead toy from the cocoa and inserted into the tins of cocoa. Smaller models were found in quarter-pound tins, while the larger ones were put into the half-pound tins, although these could sometimes contain two of the smaller cubs. The higher cost of the larger tins may account for the fact that larger cubs are more difficult to obtain than the smaller ones.

The Cococubs animals had a mascot boy leader named Jonathan of whom two versions were produced. *(see following list)*

As well as issuing the lead Cococub figures in the tins, Cadbury also started a Cococub Club, and issued an enamel members badge and a Cococub Newsletter. Two board games were produced and comic strips of the characters were printed in the daily national newspapers, family magazines, and children's comics. An annual was issued and, for the retail trade, advertisements appeared in "The Grocer," a trade magazine for shopkeepers. The amount of promotional material issued was enormous and even included cardboard cutout houses given as premiums through the Cococub Newsletter. It is not my intention to focus on these items, but to concentrate on the lead Cococub figures themselves, their dates of production, and the color variants that a collector may hope to acquire.

Solid lead Cococub figures are known to exist bearing only the Cadbury name. The origin of these solid pieces is as yet unknown. In recent years, two British collectors, Dudley Chignall and John Cupman, have continued research into the Cococub promotional scheme as a whole. Dudley has delved into the promotion in detail and has produced a most comprehensive guide to the Cadbury's Cococub scheme entitled "The Great British Book of Cococubs." I am indebted to Dudley and to John for their permission to draw upon extracts from their research and congratulate them on the wealth of information that they have accumulated.

Cadbury and Britains also worked together on several other occasions. One example of this is a small version of the Blue Bird Racing Car, Catalog #329B, which was offered via the Cococub Newsletter as one of a number of free gifts available to Cococub Club members. See the chapter on Britains Civilian Vehicles.

Britains Cococubs

	Master Mold#	Mold Made
Jonathan, Boy Eating Chocolate	673	23 March 1934
Will Mouse	654	26 March 1934
Dumpty Doo		
Peaked Cap/School Cap	653	13 April 1934
Tam O'Shanter	*See notes*	
Bowler Hat	*below on*	
Skull Cap	*head variations*	
Peter Pum Poodle	665	16 April 1934
Mrs. Henrietta Fussy Feathers	1017	18 April 1934
Mrs. Cacklegoose	648	25 May 1934
Granny Owl	1015	28 May 1934
Freddy Frog	651	29 May 1934
Silas Slink Fox	662	31 May 1934
Whiskers Rabbit	645	31 May 1934
Mr. Pie Porker	666	5 June 1934
Gussie Robin	647	5 June 1934
Mrs. Pie Porker	1016	5 June 1934
Brother Rabbit	666	8 June 1934
Piglet Pig	663	15 June 1934
Jonathan, Boy Walking	664	14 Dec. 1934

Monty Monkey	649	15 June 1935
Percy Parrot	659	18 June 1935
Bill Badger	656	26 June 1935
Squire Rooster Cockerel	667	29 June 1935
Nutty Squirrel	1018	16 July 1935
Dan Crow	655	1 October 1935
Peter Penguin	650	4 October 1935
Tiny Tusks Elephant	652	8 October 1935
Captain Kangaroo	661	18 October 1935
Timothy Tortoise	644	21 February 1936
Pat Pelican	646	5 March 1936
Tubby Bear	657	17 March 1936
Tom Kitten	658	19 March 1936

The Master Mold list was retyped by Britains towards the end of the 1930s, thus the numbering system is not consistent with the date that the mold was made.

The Smallwood Notebooks entry of April 1934 indicates; the Dumpty Doo molds have three heads/hats. It must be assumed that these were in addition to the master figure as 4 Dumpty Doo hats were issued and featured in advertisements during 1934. There could possibly be other Dumpty Doo figures with different hats. The Dumpty Doo mold was retooled on 2 November 1934 giving way to speculation that a fifth helmet hat was issued. It could be that this "helmet" is simply a poorly cast bowler hat. No specific mention is made of a Dumpty Doo with helmet in either the Smallwood Notebooks or Britains factory records.

Three different heads were obviously molded: the short, thin pointed bill head usually found wearing a skull cap; a slim, slightly open bill usually found in either school cap/peak cap or bowler hat; and thick upper bill curving upwards, beak slightly open with a shorter lower bill and found wearing a Tam O'Shanter.

Britains Smallwood Notebooks reveal that names were given to some of the Cococubs as the mold record was produced. The animals were often given other nicknames.

Ernest Aris, creator of the Cococub sketches, advertising, and promotional releases, The Cococub Games, Newsletter and various cartoon strips all from time to time printed variations of some of the Cococub names. They are as follows:

Cococub Names	**Nicknames**	**Britains Smallwood Notebooks**
Jonathon Second Boy	Jonathon (Aris)	
Will Mouse		Willie
Peter Pum Poodle	Peter Pum Dog (Aris)	
Mrs. Henrietta Fussy Feathers	Henrietta Fussy Feathers (Aris)	Fizzy Feathers
Freddy Frog	Freddie Frog (Aris)	
Silas Slink Fox	Lord Silas Slink (Cartoon)	Silas Sly
Mr. Pie Porker	Mr. Porker (Aris)	
Mrs. Pie Porker	Mrs. Porker (Aris) Pansy (Annual)	
Brother Rabbit	Rabbit Brothers (Board Game) Little Rabbit (Advert)	Binkie & Baby Rabbit
Piglet Pig	Porky Bags (Aris) The Piglets (Cococub Newsletter) Wee Peter Pig (Board Game)	Baby Porker
Squire Rooster Cockerel		Cock
Peter Penguin		Percy
Tiny Tusks Elephant	Tiny (Cococub Newsletter)	
Timothy Tortoise	Timmy (Cartoon)	
Pat Pelican	Patsy (Cartoon)	
Tubby Bear		Small Bear
Tom Kitten	Tommy (Cartoon)	

All Cococubs are courtesy of Dudley Chignall and are illustrated in order of date produced. Prices: Some color variants having been produced in lesser quantities now demand a premium in addition to the following guide.

Boy with Candy Bar. The original Jonathan. $300-$340.

Will Mouse. $55-$65.

Dumpty Doo. Skull Cap. $60-$65.

Dumpty Doo. Peak/School Cap. $60-$65.

Peter Pum. $85-$90.

Dumpty Doo. Tam 'O Shanter. $60-$65.

Dumpty Doo. Bowler. $60-$65.

Henrietta Fussy Feathers. $55-$65.

Mrs. Cacklegoose. $55-$65.

Granny Owl. $55-$65.

Freddy Frog. $55-$65.

Silas Slink. $65-$70.

Whiskers Rabbit. $85-$90.

Gussie Robin. $55-$65.

Mr. Pie Porker. $75-$85.

Brother Rabbit. $55-$65.

Mrs. Pie Porker. $75-$85.

Piglet Pig. $55-$65.

Piglet Pig. $55-$65.

Percy Parrot. $55-$65.

Bill Badger. $55-$65

Boy Jonathan. $220-$250.

Monty Monkey. $55-$65.

Squire Rooster. $85-$90.

Nutty Squirrel. $55-$65.

Tiny Tusks. $85-$90.

Dan Crow. The headless figure was sold as part of the Britains Archive Auction Lot at Christie's, London in 1994. $55-$65.

Captain Kangaroo. $65-$75.

Peter Penguin. $65-$75.

Timothy Tortoise, front. $60-$70.

Timothy Tortoise, back. $60-$70.

Tom Kitten. $60-$70.

Dumpty Doo. Uncataloged hat (? helmet). Unique.

Pat Pelican. $60-$70.

Tubby Bear. $65-$70.

Cocucubs in production sequence.

Original Box with Original 15 Cubs. Sample set sent by Britains to Cadbury for approval. Unique.

Cadbury's Cocoa Tins which included the Cubs. $300-$350 each.

Badge with Veterans talisman. $60-$65.

Original Artist's Impressions of Cubs by Ernst Aris, courtesy of Arnold Rolak.

Artist's impression of Peter Pum Dog.

Artist's impression of Original Boy Jonathan.

Artist's impression of Willie Mouse.

Top left: Artist's impression of Whiskers Rabbit.

Top center: Artist's impression of Henrietta Fussy Feathers.

Top right: Artist's impression of Dumpty Doo.

Bottom center: Artist's impression of Silas Slink Fox.

Bottom right: Artist's impression of Mrs. Cacklegoose.

Top left: Artist's impression of Freddie Frog.

Top center: Artist's impression of Granny Owl.

Top right: Artist's impression of Mrs. Porker. Later renamed Mrs. Pie Porker.

Bottom left: Artist's impression of Mr. Porker. Later renamed Mr. Pie Porker.

Bottom center: Artist's impression of Gussie Robin.

Bottom right: Artist's impression of Porky Bags. Later renamed Peter Pig.

23. Pond and Christmas, Mid-1930s

During the 1930s, most of the production at Britain's three factories in London, Liverpool, and Stoke was taken up with the additions to the catalog of many items for other manufacturers. These included souvenirs, novelties, advertising and promotional items ... anything in fact that would bring additional revenue to the company during the Depression years. The items listed above have been itemized in their respective chapters.

This chapter shows a number of ranges produced from 1932 to 1941 that for reasons of size and rarity have been grouped together.

Christmas Novelties. Factory Sample Card containing what is thought to be the full range of Christmas items shown in catalogs beginning in 1932. Top row, left to right: #2X Father Christmas, small; #3X Father Christmas, large; #7X Robin, small; #8X Robin, large (both the #7X and #8X Robins are of a plaster composition material and have wire feet); #14X Robin on spring with log and nest and eggs, large. Second row, left to right: #20X; #3X; #13X; #21X and #22X Robins on logs. All fit into the log by pin or spring. Third row, left to right: #47X Stag and #40X Gnome. Bottom row, left to right: #25X Baby Boy waving, four color variants; #26X Baby boys standing and sitting on sleigh; #39X Large Father Christmas pencil sharpener (see figures made for other companies chapter); #31X Teddy Bears, four color variants (this item is a paint adaptation of the seated bear cub from the Zoo Series); #37X Reindeer (not illustrated). Right, top to bottom: #41X, #42X, #43X. Three sizes of Robins with wire feet (this item was originally offered in the Christie's Britains Archive Sale in 1994 and is unique). *Courtesy Christie's Auction House, London/The Arnold Rolak Reference Collection.* $3,000-$4,000.

Garden Ornaments Sample Card. All figures are large scale. Top row, left to right: #526B Yellow Wagtail standing; #563B Yellow Wagtail feeding; #564B Bull Finch. Second row, left to right: #565B Bull Finch feeding; #566B Red Breasted Finch standing; #567B Red Breasted Finch Feeding. Bottom row, left to right: #568B Mushroom; #569B Toadstool; and #614B Snail. *Courtesy Christie's Auction House, London/The Arnold Rolak Reference Collection.* $3,000-$3,500.

Garden Gnomes: left to right. #J1 Large Garden Gnome with pipe; #243B Small Gnome lying; #242B Large Gnome lying; and #J5. Large Garden Gnome standing with flower pot. A large standing Gnome with hands in pockets was allocated #242B pre-War and #J4 post-War (not illustrated). "J" prefixes are post-War numbers. *Courtesy Christie's Auction House, London/The Arnold Rolak Reference Collection.* Left to right: $100-$150, $60-$70, $100-$150, $100-$150.

Garden Animals: #483B Frog and #485B Newt. Both were produced in hollow or solid cast form. *Courtesy Christie's Auction House, London/The Arnold Rolak Reference Collection.* $80-$100 each.

Garden Gnomes, small. Left and right: two examples of #169B Gnome standing. Second left: #243B Gnome lying. Center: #487B very small Gnome, a paint variant of #40X from the Christmas Series, #170B Gnome sitting, hands on knees; #171B Gnome seated, cross legged (not illustrated). *Courtesy Christie's Auction House, London/The Arnold Rolak Reference Collection.* $60-$70 each.

A selection of items already illustrated and described, including the only known version of the Ladybird, part of the Garden/Pond Series. *Courtesy Christie's Auction House, London/The Arnold Rolak Reference Collection.* Ladybird $500-$600. Others $300-$350.

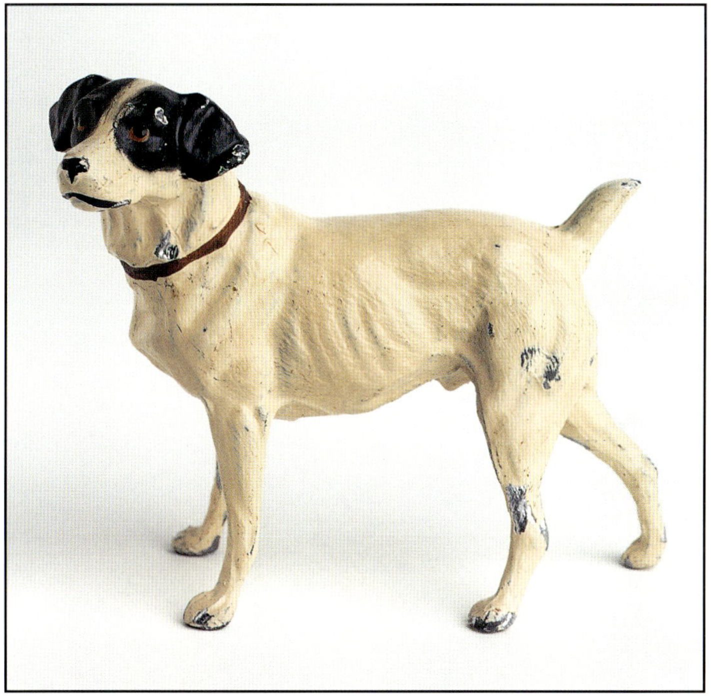

Terrier, Catalog #7H. A Pug dog was also made, but is exceptionally rare, allocated catalog #9H (not illustrated). *Courtesy Christie's Auction House, London/The Arnold Rolak Reference Collection.* $300-$350.

Stag #10H and Doe #11H. *Courtesy Christie's Auction House, London/The Arnold Rolak Reference Collection.* $180-$220 each.

Seagull wings outstretched #12H, Seagull seated #22H. $40-$60 each.

24. Games and Sports, Mid-1930s

The mid-1930s saw the issue of many souvenir, novelty, promotional, and character items, as well as Sports and Games related items, many made by Britains for other companies. With the exception of the Great Test Match Game and Chess set, the following items came from this thirties period.

Cricketer and Engine Driver #809. The Cricketer paint variant is believed to date from around 1910. Cricketer $160-$200.

The Great Test Match County Championship. A set of eleven Cricketers issued by a company named Pealland and numbered 115. This manufacturer of board games commissioned Britains to produce these Cricketers which were adapted from the railway driver or stoker figure. See next illustration. *Courtesy Christie's Auction House, London/The Arnold Rolak Reference Collection*. $3,500-$4,000.

Cricketers, two full teams, twenty-two in all. *Courtesy Alf Mussett*. $160-$200 each Cricketer.

Shamrock III, America's Cup Challenge Yacht and two smaller yachts #30D flat items made for board games possibly for Frederick Warne. *Courtesy Christie's Auction House, London/The Arnold Rolak Reference Collection*. $80-$100.

Motor Racing Car game pieces (see previous illustration). Large Car #29D, small car #33D. Racing Cyclists #31D were also manufactured (not illustrated). See Christie's Britains Archive Catalog 1994. *Courtesy Christie's Auction House, London/The Arnold Rolak Reference Collection*. $40-$60 each.

Cricketers with paint variants. The green hat is believed to represent an Australian Cricketer. *Courtesy Philip Dean Reference Collection*. Green and blue hat variants, $250-$280 each.

Jockey and Racehorse #123B issued in 1932 for racing game and consistent in scale with 54mm. Miniature racehorse and jockey #19D. *Courtesy Christie's Auction House, London/The Arnold Rolak Reference Collection.* $40-$60 each.

Speedex Racing Game employing #123B racehorse and jockey galloping. *Courtesy Peter Kirk.* $300-$400.

Greenhill/Courtenay galloping racehorse prototype similar to those manufactured by Britains.

Set #1338 Miniature Golf. This extremely rare set was introduced in 1934. Two golfers with two sizes of bunkers. Each golfer's arms and head are moveable. Set #1337 offered only one golfer (not illustrated). *Courtesy Christie's Auction House, London/The Arnold Rolak Reference Collection.* $4,500-$5,000.

Set #1344 Miniature Archery set in original box. Another 1934 release, the Archer's head is moveable and the wire bow will fire the arrows included. Set #1340 (not illustrated) was a larger set containing 2 archers. *Courtesy Christie's Auction House, London/The Arnold Rolak Reference Collection.* $4,000-$4,500.

Set #1344 Miniature Archery set in original box with additional paper target. *Courtesy Christie's Auction House, London/The Arnold Rolak Reference Collection.* $4,200-$4,500.

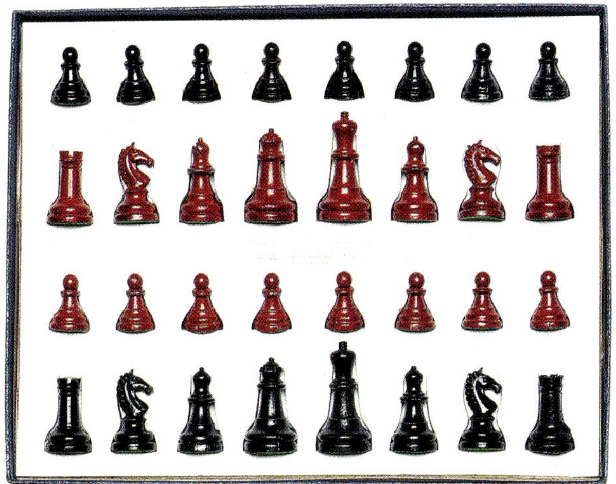

Set #1920 Chess set in original box. Master Model numbers: Pawn, red #1113B, black #1119B; Castle, red #1114B, black #1120B; Knight, red #1115B, black #1121B; Bishop, red #1116B, black #1122B; Queen, red #1117B, black #1123B, King, red #1118B, black #1124B. *Courtesy Christie's Auction House, London/The Arnold Rolak Reference Collection.* $350-$450.

Set #1344 Miniature Archery set box art close up. *Courtesy Christie's Auction House, London/The Arnold Rolak Reference Collection.*

25. Circus, 1936

The Mammoth Circus Range was introduced in 1936 and continued post-War. The following Pre-War issues were discontinued and were not issued after the commencement of production post-War: Catalog #1439 Circus Roundabout;, Catalog #1441 Flying Trapeze; Catalog #1444 Mammoth Circus (vertical) display set; and the single item of Cowboy with Rope #356B. Catalog #2054 was the only post-War addition, a display set containing a combination of the standard Circus Range items. Production of all circus items ceased after 1961.

Set #2054 Circus display with twelve pieces was introduced post-War (not illustrated). Circus items were also offered as second grade or reduced painting. These items are very rare. Some were offered on paint sample cards in the Britains Archive Sale. Master Model Numbers were allocated as follows. See Christie's Auction/Britains Archive Paint Sample Cards, 1994:

#604B	Clown in Baggy Pants
#605B	Clown with Hoop
#606B	Boxing Clown
#607B	Equestrienne
#608B	Lion Tamer
#609B	Clown on Stilts
#610B	Ringmaster

Circus Advertisement

#351B Prancing Liberty Horse. $65-$75.

#354B Clown with hoop. $75-$80.

#355B Equestrienne. Can be used with #352B Trotting Horse or #359B Walking Elephant. The small stand used to support both this and #354B Clown with Hoop was also used as a garden earth item. Circus Master Model #1212. $55-$60.

#352B Trotting Horse. Can be used with #354B Clown with Hoop or #355B Equestrienne. $60-$65.

#353B Man on Stilts. Britains Factory Records list this item as a boy. $70-$75.

#355B Equestrienne. Pink color variant. $65-$70.

#356B Cowboy with Lasso. Pre-War issue only. Base of figure is sand-colored as opposed to the normal green used when this item appeared in the Wild West Series. *Courtesy Philip Dean Reference Collection*. $150-$200.

#359B Elephant Walking. Can be used with #354B Clown with Hoop and #355B Equestrienne. $65-$75.

#466B Circus Tub. Can be used with #448B Performing Tiger or #449B Performing Elephant. $65-$70.

#357B Ring Master. $50-$60.

#358B Clown standing wearing baggy pants. $55-$65.

#447B Boxing Clown. $65-$70.

#448B Lion Tamer. $50-$55.

#451B Boxing Kangaroo. $65-$70.

#449B Performing Tiger. Post-War only issue in Circus Series. $40-$45.

#451B and #447B Boxing Kangaroo and Clown. $120-$130 pair.

#450B Performing Elephant. $80-$90.

#447B and #451B Boxing Clown and Boxing Kangaroo. $120-$130 pair.

#354B and #359B Clown with Hoop and Elephant. $140-$150.

#355B and #359B Equestrienne and Elephant. $125-$130.

Artist's impression of Equestrienne and Trotting Horse not adopted by Britains. *Courtesy of Arnold Rolak.*

Below: Set #1439 Circus Roundabout. Pre-War only issue. The Roundabout was allocated Master Model #976. Master Mold #98 was allocated to the Roundabout boy on a horse and #99 to girl on a horse. *Courtesy Christie's Auction House, London/The Arnold Rolak Reference Collection.* $3,000-$3,500.

#352B and #355B Trotting Horse and Equestrienne. $120-$130.

Circus display with wooden circus ring. Ring allocated Master Model #384B.

Set #1441 Flying Trapeze. Pre-War only issue. Master Model #977 was allocated to the trapeze. Master Mold #546 was allocated to seated girl for trapeze and #547 to the trapeze clown. *Courtesy Philip Dean Reference Collection.* $1,000-$1,800.

Set #1443 Mammoth Circus. Showing paint variations. Pre-War issue only. *Courtesy Philip Dean Reference Collection.* $600-$800.

Set #1444 Circus display box art. *Courtesy Philip Dean Reference Collection.*

Set #1444 Circus display. Pre-War issue only. Items are designed to stand in box when lid is removed. *Courtesy Philip Dean Reference Collection.* $2,500-$3,000.

Set #1442 Mammoth Circus. Ringmaster, Man on Stilts, Clown in baggy pants, Prancing Horse, Trotting Horse, Equestrienne, and stand. Pre-War issue only. *Courtesy Philip Dean Reference Collection.* $600-$800.

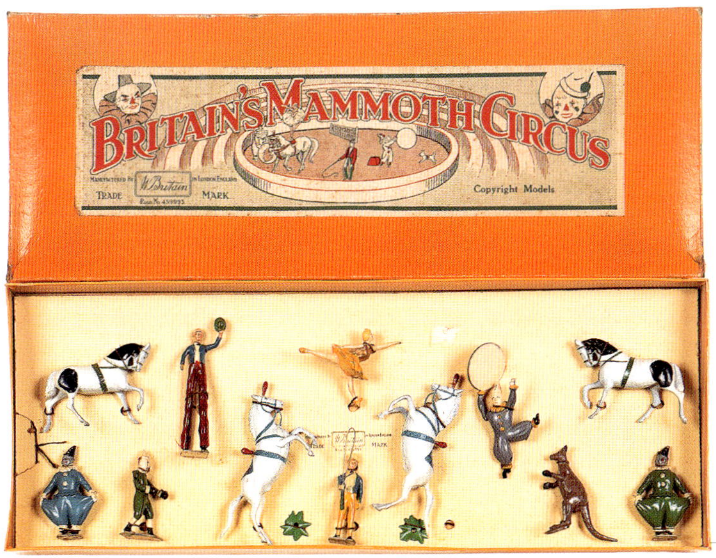

Set #1443 Mammoth Circus. Pre-War issue only. *Courtesy Philip Dean Reference Collection.* $600-$800.

Set #1539 Large Circus display. Pre and post-War issue. $1,200-$1,500.

26. Lilliput, Post-War

During the early post-War period, Britains formed an association with W. Horton ("Toys and Games"), Ltd., a company based in Cleveland. Horton was responsible for the Lilliput range of figures. The standard-size Farm figures were pantographed down in size while Railway figures and vehicles of a completely new design were manufactured. All were to fit the Double "O" Gauge Railway scale which became the standard toy train scale in England after the war. The range was probably introduced in 1954. The average size of a standing figure is 21mm with vehicles and animals in proportion. Unless otherwise stated, the illustrations in this chapter are from the Philip Dean Reference Collection.

Lilliput World Catalog cover.

Lilliput World Catalog cover.

Left to right: #LB514, Shire Horse; #LB515, Farmer; #LB516, Farmer's Wife; and #LB517, Nurse with Child. The latter item was not produced in 54mm. Left and right: $2-$3. Others, $5-$6.

Lilliput Series point of sales advertising artwork.

LILLIPUT SERIES United Kingdom Public Selling Price including Purchase Tax

No.	Item	Price
LB/514	Shire Horse	–/9½
LB/515	Farmer	–/9
LB/516	Farmer's Wife	–/9
LB/517	Nurse & Child	/11½
LB/518	Foal	–/4½
LB/519	Cob	–/7½
LB/520	Standing Cow	–/8½
LB/521	Feeding Cow	–/8½
LB/522	Calf Lying	–/4½
LB/523	Collie Dog	–/3½
LB/524	Goose	–/3½
LB/525	Sheep, standing	–/4
LB/526	Sheep, feeding	–/4
LB/527	Pig	–/3½
LB/528	Lamb	–/3½
LB/529	Ducks & Drakes	–/3½
LB/530	Hurdle	–/2½
LB/531	Stable Lad	–/9
LB/532	Land Girl	–/9
LB/533	Porter with Trolley	–/10
LB/534	Guard	–/7½
LB/535	Station Master	–/7½
LB/536	Civilian Motorcyclist	1/1½
LB/537	Porter with Luggage	–/9
LB/538	Newsvendor	–/7½
LB/539	Lady with case	–/7½
LB/540	Man with book	–/7½
LB/541	Man with umbrella	–/7½
LB/542	Lady with hatbox	–/7½
LB/543	Golfer	–/7½
LB/544	Barrel	–/4
LB/545	Hamper	–/4
LB/546	Large Packing case	–/4
LB/547	Small Packing case	–/4
LB/548	Telegraph Boy	1/1½
LB/549	Four wheel trolley	–/4½
LB/550	Speed Cop	1/1½
LB/559	Huntsman, Top Hat	1/1½
LB/560	Huntswoman, Top Hat	1/1½
LB/561	Hounds, legs out-stretched	–/4½
LB/562	Hounds, legs closed	–/4½
LB/563	Fox	–/3½
LB/564	Huntswoman, Bowler	1/1½
LB/565	Huntsman, Bowler	1/1½
*LB/567	Infantry, charging	–/5
*LB/568	Infantry, wounded	–/5
*LB/569	Radio operator	–/5
*LB/570	Infantry, throwing grenade	–/5
*LB/571	Infantry, standing firing	–/5
*LB/572	Infantry, kneeling firing	–/5
*LB/573	Infantry, attacking	–/5
*LB/574	Officer, with pistol	–/5
VEHICLES		
LV/601	Open Sports Car	1/10
LV/602	Saloon Car	1/10
LV/603	Articulated Lorry	2/6
LV/604	'Fordson' Tractor	2/6
LV/605	Milk Float	2/–
LV/606	Tumbrel Cart	2/–
LV/607	Army covered 3-ton Truck	3/4½
LV/608	Civilian 3-ton Truck	2/10
LV/609	Austin 'Champ'	2/9
LV/610	Centurion Tank	4/11
LV/611	Self-propelled Gun	4/3
LV/612	Army 1½-ton Truck	2/10
LV/613	Army covered 1½-ton Truck	3/4
LV/614	Articulated Truck	3/9
LV/615	'Saracen' Personnel Carrier	2/10
LV/616	Civilian 1½-ton Truck	2/10
LV/617	Civilian Ambulance	3/3
LV/618	Army Ambulance	3/3
LV/619	P.O. Van	3/3
LV/620	Army 3-ton Truck	2/10

**Plastic*

Lilliput price list giving full listing of items in the range.

Left to right: #LB518, Foal; #LB519, Cob; #LB520, Standing Cow; #LB521, Feeding Cow; and #LB522, Calf Lying. $2-$3 each.

Left to right: #LB523, Collie Dog; #LB524, Goose; #LB525, Sheep Standing; #LB526, Sheep Feeding; and #LB527, Pig. $2-$3 each.

Left to right: #LB528, Lamb; #LB529, Duck/Drake; #LB530, Hurdle; #LB531, Stable Lad; and #LB532, Land Girl. Left to right: $2-$3, $2-$3, $2-$3, $5-$6, $5-$6.

Left to right: #LB533, Porter with Barrow; #LB534, Guard; #LB535, Station Master; #LB536, Civilian Motorcyclist (not illustrated); and #LB537, Porter carrying Luggage. These figures were not produced in 54mm. $5-$6 each.

Railway Station Accessories. Left to right: #LB544, Barrel; #LB545, Hamper; #LB546, Large Packing Case; #LB547, Small Packing Case; and #LB549, Electric Trolley. #LB548, Telegraph Boy on Motorcycle and #LB550, Speed Cop are illustrated in Set #L/11. These figures were not produced in 54mm. Accessories $4-$5 each. Motorcycles $20-$25 each.

Left to right: #LB538, News Vendor; #LB539, Lady with Case; #LB540, Man with Book; #LB541, Man with Umbrella; #LB542, Lady with Hat Box; and #LB543, Golfer. These figures were not issued in 54mm. $5-$6 each.

Left to right: #LB559, Huntsman in top hat on galloping horse; #LB560, Huntswoman in top hat on galloping horse; #LB561, Hound legs outstretched (not illustrated); #LB562, Hound running legs closed; #LB563, Fox running; #LB564, Huntswoman in bowler hat on galloping horse; and #LB565, Huntsman in jockey cap on galloping horse. Huntsmen, $10-$12 each. Dogs, $4-$5 each.

Gaps in numbering systems indicate items were military vehicles. See Christie's Auction/Britains Archive 1994 illustration.

#LV601. Open Sports Car. *Courtesy Christie's Auction House, London/The Arnold Rolak Reference Collection*. $75-$85 each.

#LV602. Saloon Car. *Courtesy Christie's Auction House, London/The Arnold Rolak Reference Collection*. $75-$85 each.

#LV601. Open Sports Car with second version box. $120-$140.

#LV601. Open Sports Car with original box. $100-$120.

#LV603. Articulated Lorry. *Courtesy Christie's Auction House, London/ The Arnold Rolak Reference Collection*. $85-$90 each.

#LV604. Fordson Tractor with second version original box. $120-$140.

#LV603. Articulated Lorry with second version box. $140-$150.

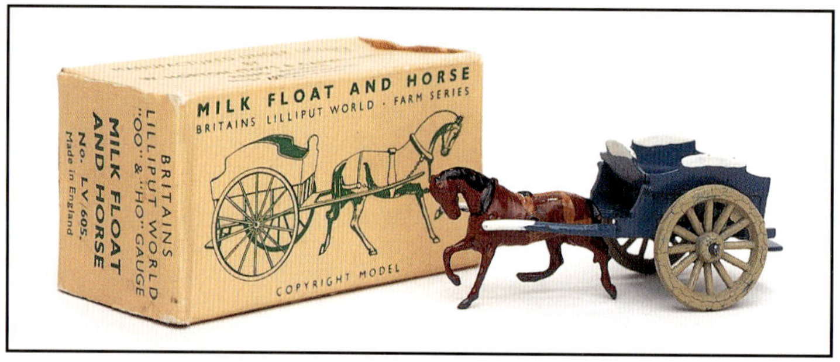

#LV605. Milk Float with original box. $100-$120.

#LV605. Milk Float, second version box. $120-$150.

#LV604. Fordson Tractor with original box. $100-$120.

#LV609. Austin Champ. Although in theory a military vehicle, this item was included as the civilian version of the Austin Champ in Set #L/11. See illustration. $70-$75.

#LV606. Tumbrel Cart with original box. $100-$120.

#LV606. Tumbrel Cart, second version box. $120-$150.

#LV614. Articulated Truck, high sided. *Courtesy Christie's Auction House, London/The Arnold Rolak Reference Collection*. $85-$90.

#LV608. Civilian 3-ton truck. *Courtesy Christie's Auction House, London/The Arnold Rolak Reference Collection*. $80-$85.

#LV616. Civilian/Farm 1-1/2 ton truck. $80-$90.

#LV616. Civilian/Farm 1-1/2 ton truck. Paint variants. $80-$90 each.

#LV617. Civilian Ambulance. $100-$110.

#LV616. Civilian/Farm 1-1/2 ton truck with second version box. $180-$220.

#LV619. Royal Mail Van. $120-$130.

#L/7. Farm display box. *Courtesy Christie's Auction House, London/The Arnold Rolak Reference Collection.* $300-$400.

#L/9. Country Cottage with original box. *Courtesy Stewart Saxe.* $100-$120.

#L/10. Barn or Cart Stable with original box. *Courtesy Stewart Saxe.* $100-$120.

#L/8. Roadside Inn or Country House. $60-$80.

#L/11. Railway Personnel and Civilian Vehicles. $500-$600.

#L/11. Variation which includes two non-cataloged "A.A." boxes. $500-$600.

#3/128. Retail counter display box containing Farm and Railway items. *Courtesy Christie's Auction House, London/The Arnold Rolak Reference Collection.* $400-$450.

"A.A." boxes possibly made by another company for inclusion into Set #L/11. $35-$45 each.

#3/135. Retail counter display box containing Farm figures. *Courtesy Christie's Auction House, London/The Arnold Rolak Reference Collection.* $350-$400.

#3/137. Retail incomplete counter display box containing Hunt figures. *Courtesy Christie's Auction House, London/The Arnold Rolak Reference Collection.* $250-$280.

Display boxes with catalog and price list for Farm Series. *Courtesy Christie's Auction House, London/The Arnold Rolak Reference Collection.* Small $80-$100. Medium $200-$250. Large $300-$400.

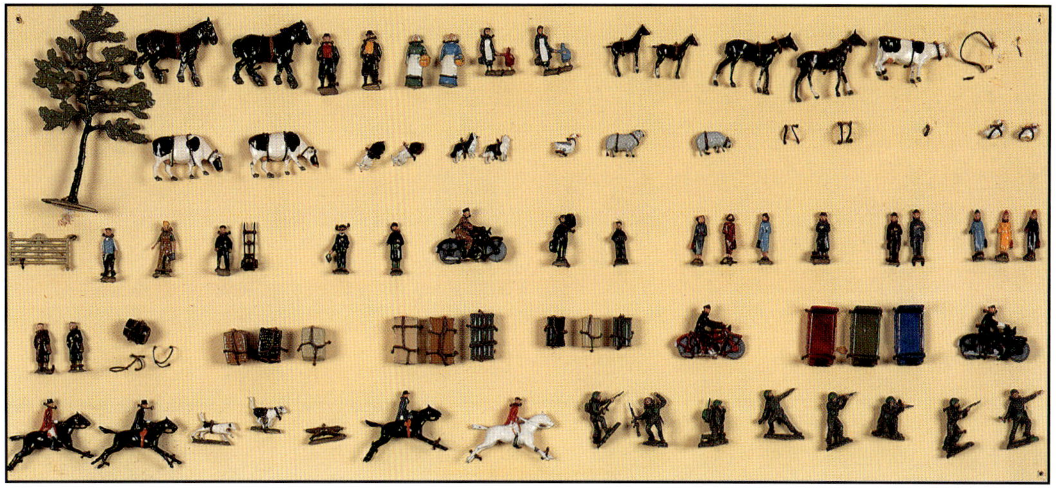

Factory sample card with examples of the entire Lilliput range. *Courtesy Christie's Auction House, London/Britains Archive.* Unique.

Prototype color variations of vehicles from Britains Archive. *Courtesy Christie's Auction House, London 1994.* Unique.

Rare salesman's sample containing one of each figure from Railway and Farm range. *Courtesy Philip Dean Reference Collection.* $800-$1,000.

A typical scene using Lilliput cottages and figures as used in 00/H0 Railway layouts.

BIBLIOGRAPHY

Joplin, Norman. *British Toy Figures*. Arms and Armour Press, 1987.
_____. *The Great Book of Hollow Cast Figures*. New Cavendish Books, 1993.
Opie, James. *The Great Book of Britains*. New Cavendish Books, 1993.
Kurtz, Henry and Burtt R. Erlich *The Art of the Toy Soldier*. New Cavendish Books, 1987.
Fawdry, Marguerite. *British Tin Toys*. New Cavendish Books, 1990.
Presslands Great Book of Tin Toys. New Cavendish Books, 1999-2000.
Britains Archive Auction Catalog. Christie's Auction House, 1994.

The following are all privately published by the authors:

Chignall, Dudley. *The Great British Book of Cococubs*. 2000.
Kirk, Peter. *Britains Racing Colors of Famous Owners*. 2001.
Wallis, Joe. *Armies of the World*. 1993.
Wallis, Joe. *Regiments of All Nations*. 1982, Revised 1992.

All of the above include information on Britains Non-Military Items.

Other books by Norman Joplin:

American Dimestore Toy Soldiers and Figures. Schiffer Publishing, 2000. With Don Pielin and Verne Johnson.
Toy Soldiers. Apple Press, UK Edition, 1994.
Toy Soldiers. Courage Books, US Edition, 1994.
Identifying Toy Soldiers. Quintet Publishing, 1996. Also published in German, 1994.
Toy Soldiers. Grange Books, Collectors Corner Edition, 1999.